This Cold House

THIS COLD HOUSE

The Simple Science of Energy Efficiency

Colin Smith

THE JOHNS HOPKINS UNIVERSITY PRESS
Baltimore

© 2007 The Johns Hopkins University Press
All rights reserved. Published 2007
Printed in the United States of America on acid-free paper
9 8 7 6 5 4 3 2 1

The Johns Hopkins University Press
2715 North Charles Street
Baltimore, Maryland 21218-4363
www.press.jhu.edu

Library of Congress Cataloging-in-Publication Data

Smith, Colin, 1942–
    This cold house : the simple science of energy efficiency / Colin Smith.
    p.   cm.
    Includes index.
    ISBN-13: 978-0-8018-8622-5 (hardcover : alk. paper)
    ISBN-10: 0-8018-8622-8 (hardcover : alk. paper)
    1. Dwellings—Energy conservation. 2. Dwellings—Heating and ventilation. I. Title.
TJ163.5.D86S635 2007
    697—dc22   2006033658

Contents

Preface

I suppose that all my life I've been thought of as a bit of an oddball. But that ain't so bad here in the UK where I live. We have a strong tradition of eccentricity, honoring our nonconformists, and amusing the rest of the world by our careless antics.

My eccentricity surfaced when I found I was not only good at but thoroughly enjoyed science and math, particularly physics. This was back in the mid-fifties, just as rock and roll surfaced, and Elvis took the world by storm. Even then physics was thought of as a "hard" subject. Everyone had to study it because it was needed for so many careers, but there was definitely something weird about actually liking it.

In due course I took a physics degree and went on to teach the subject I loved, eventually lecturing at the Manchester Metropolitan University. But not in pure physics—there are too few students for that these days on this side of the pond. My special interests were to do with energy, especially energy-saving devices used in the home, and this fits nicely into a number of courses.

The second strand of my eccentricity came from my interest in environmental issues. This began in the early seventies just as Greenpeace and Friends of the Earth came on the scene. Although I have never been a member of either, they certainly put environmentalism on everyone's agenda, including mine. Some of the predictions made at the time, however, turned out to be spectacularly wrong. I well remember predictions of oil running out in forty years. We are nearly there, and it ain't happened yet! But it was this that sparked my interest and concern.

This book combines the two strands of my interests applied to a particular problem—that of how to create warm and comfortable homes from "this cold house," in ways that are kind to the earth. But it is not a "deep green" approach. There are plenty of specialist books that meet this need. No, rather it is my conviction that we can all play a significant part in tackling environmental problems—by each making small, realistic, and achievable adjustments to our lives.

It is my aim to inform your decision making about what to buy and how to use it to keep comfortable in an economic and environmentally friendly way. This is not so much a "how-to book" as one that helps you make an educated choice about what is needed in your particular circumstances. It could well save you from plunging in and making an unwise investment. It will certainly show how most people can make fuel and cost savings in their homes.

This book is based on my teaching. It is not a textbook, but it is useful as a preliminary reader that effectively supports the more dry academic approach often found in standard texts. Explanations are given in detail and the book is illustrated with a wealth of anecdotes unlikely to be found in a textbook. Think of it as a stimulating companion to such texts for students of subjects such as architecture, building science, home economics, and environmental health.

My writing style may come as a surprise—but I hope it is a pleasant one. Be prepared to be introduced to Grandma, the school janitor, and a whole host of my relatives from before World War II. Their struggle to achieve warmth and comfort in their homes against a background of fuel scarcity and economic deprivation exemplifies many of the points of this book. You may also be surprised to find details of the historical development of much of the equipment available to us. Such detail eases the explanations and forms an interesting backdrop to the workings of the vital but often hidden paraphernalia so essential to our comfort.

I am indebted to Trevor Lipscombe, my editor, for encouraging me to write in this style. I hope that you find the anecdotes and historical detail fascinating and amusing. The intention is that they illustrate points, making the book readable and informative.

Enjoy it!

This Cold House

Finding Your Way

It was through my grandpa that I first became aware of the phrase "The good old days." For him, a working-class Brit, those were the early days of the last century. The straight-laced and morose Queen Victoria—the "Old Queen"—had died, and the fun-loving playboy king Edward VII was on the British throne. It was the heyday of the music hall (from which the variety show later developed). And what's more, you could leave your door unlocked at night, and everyone cared about and generally looked after everyone else. At least that is how my grandfather remembered it—the reality may well have been different.

For me, born in Great Britain in the middle of World War II, my early days were anything but good. One of my earliest memories was of the Great Freeze of 1947, when the snow reached the roofs of houses, and winter seemed to last for ever. To make matters worse, a national fuel shortage sent Dad "coal picking" around the yards and spoil heaps of our local coal mines.

Indeed, the luxury of all-year-round warmth was not achieved by our family until well into the era of Elvis and the Beatles. Like many others at the time, I grew up to the castanet sound of chattering teeth and slept in a bedroom that felt like the inside of a freezer. There was ice on the windows and we ate huddled around the parlor fire, from where we rarely moved.

Nowadays, warmth and comfort are taken for granted. But when our heating breaks down, the resulting discomfort hits us with a vengeance—we shiver, lose concentration, and generally feel miserable. The effects of thermal discomfort on our bodies are not pleasant.

And thermal discomfort works both ways. Had you been in Dar es Salaam, Tanzania (on the east coast of Africa near the equator), for Christmas 1965, you may well have found me and Gerry, my American Peace Corps buddy, wandering, somewhat aimlessly, around the few air-conditioned shops in the city. Young teachers seeing a bit of the world before we settled down, we both wound up in the same school. Neither he (from Seattle) nor I (from England) had realized just how unbearable the tropics can be at that time of year. I still don't know how we managed to evade heat stroke.

Recalling events and experiences like these remind us that current standards of comfort have only recently become widespread. The development and availability of an enormous range of appropriate, affordable equipment has been the crucial factor in achieving this advance.

Yet there is a problem. Since the early 1970s, the realization has grown that much of the fuel that gives us our comfort comes from nonrenewable resources—fossil fuels (oil, gas, and coal) laid down in the earth's crust over geological time scales. Once they are used up, they can only be replaced over similar time scales—and we cannot wait that long for them to be replenished. What's more, when they burn they release gases such as carbon dioxide that contribute to global warming and climate change—another potential catastrophe in the offing.

The environmental impacts of such an apparently simple activity as keeping comfortable are potentially large and should be the concern of all responsible consumers. The challenge is to provide acceptable warmth and comfort with minimum environmental impact and at affordable costs.

This book aims to help meet the challenge by supplying the necessary knowledge in an entertaining and interesting way. It describes and assesses equipment and systems in terms of effectiveness, running costs, environmental impacts, and other factors. Yet any equipment is only as good as the person operating it, for mismanaged heating and air-conditioning equipment is wasteful—both in terms of fuel and dollars. Therefore, ways of optimizing the use of the equipment are also considered.

Your Map

My home is on the doorstep of the Peak District (www.visitpeakdistrict.com) in the UK, and it is easy for me to "chill out" by walking on the surrounding hills and moors. On the moors, time stands still. Yet, they are killers of the unwary. Our mountain rescue team is out virtually every weekend to bring down someone who has gotten into trouble—generally through inadequate preparation. Sadly, a few walkers die up there each year, often getting lost as mists suddenly come down to obscure the view.

All of this could be avoided if the hikers had learned how to use a map and compass—a skill that I learned at the age of 13. Dad saw that I was interested in the great outdoors and sent me on a hillcraft training course to learn how to enjoy the hills safely. Since then I have always liked to know exactly where I am and where I'm going and to have a plan for getting there.

The chapters of this book follow a logical sequence, and the following short overviews give you a 'map' to help you find your way around. I start in Chapter 1 with a discussion of comfort. At first this might seem easy—just ensure that the temperature around us is right. But it is not as simple as this. Our bodies respond differently to different types of heat, and the balance between them needs to be right. Dampness is another major factor. Dar es Salaam was hot, but what really got to us was the combination of the heat and the high humidity from the sea. Our sweat just didn't evaporate, which made us very uncomfortable—and smelly!

How I wished we had an effective heating system in 1947. At night we wore sweaters, old trousers, and thick socks—even in bed. On particularly cold nights, Mom would cover the beds with our outdoor clothes as well as with every blanket she could find. Heat was lost from the bed very slowly with all that insulation over us, and it was amply replaced by the heat of our own bodies. In Chapters 2 and 3, I consider ways our homes, like our bodies, can be "wrapped up warm," so that the heating system can cope with the cold at an economical cost and with minimal environmental impact.

Shivering helps us keep warm and sweating cools us down. In both situations we use energy, and the fuel that supplies the energy is the food we eat. Our household systems need fuel too, and traditionally these have been the fossil fuels, either burned directly at home or first converted into electricity by being burned at the power plant. In Chapter 4, I compare all the readily available types of fuel in terms of costs, safety, environmental impacts, and convenience in use.

The long hours sitting around our parlor fire molded us together as a family. We gossiped about the latest family and local news, we put the world right, and we listened intently to the comedy and entertainment shows on the radio. Even today, when central heating (or cooling) systems have come of age, there is something missing in a home without a fire and fireplace. Ours was little more than hot coals resting on a grate under a chimney, but nowadays there are many different, attractive designs available. These are described, categorized, and compared in Chapter 5.

Eventually we upgraded to a rather basic central heating system, and now as my sons start to set up their homes, even I am astounded at the variety of heating and air-conditioning systems available. When faced with this kind of situation, it helps to categorize the systems into types—it's much easier to deal with a small number of categories than a large number of individual items. Fortunately, heating systems divide easily and naturally into only four types—warm air, steam, hydronic, and electrical. Air-conditioning systems are all technically similar, forming a further category on their own. These are described and compared in Chapter 6.

A major problem that both Grandpa and I had in our respective "good old days" was that of control. If we were too cold, we put on extra clothes and lit a fire. If we were too hot, we took some clothes off and damped down the fire. Today we have thermostats, humidistats, and timers to keep our home comfortable for us. In Chapter 7, I describe and compare control equipment and consider the settings that might best achieve comfort at low economic and environmental costs. I also show how to estimate an annual fuel bill.

In Chapter 8, I indulge in a little futurology by considering current innovations that could well become mainstream. Of particular interest is the possibility of *microgeneration*—generating your own power at home—a field in which there are many interesting and promising developments. The necessary equipment currently tends to be more costly than conventional systems, but who knows, we could all wind up using it earlier than we now think!

Finally, the epilogue rounds off the book with an overview of our journey.

Your Plan

The above constitutes your map for finding your way around this book, but you may need a reading plan to act as a compass as well. Here are some suggestions.

One viable plan is to read right through from start to finish. No assumptions are made about prior knowledge, and the book has been carefully structured to make it easy to follow. Furthermore, human interest stories are included to enhance its appeal—the book is intended to be enjoyable as well as informative.

Alternatively, you may have been attracted to buy the book because you are considering a major improvement to your heating or air-conditioning system. Dipping into it to find out more before committing yourself to a large investment is another workable plan. Each chapter can be read independently of the rest—though this may mean occasional glancing back to an earlier section. The index should enable you to quickly find information of interest.

I spend all of my working life teaching, the bulk of it at a university where the subject matter of this book formed part of a wide range of different courses. Throughout that time I was unable to find a readable introductory text to inspire my students. A good plan for freshmen students considering a career as heating engineers (or architects or home economists) is to first read through this book to get a reliable overview and then use it later for reference.

Whatever your plan, puzzling technical terms may crop up from time to time. These are explained the first time that they are used and collected together in a glossary at the end of the book. More problematic is that some ordinary words, such as *heat, energy,* and *work,* also have special technical meanings. To most of us, *heat* is simply "that which makes things hot," although the high school physics student may know it also as "thermal energy." Yet both these are strictly incorrect; for physicists would insist that the true definition has to do with energy on the move. Nonetheless, I, like many authors, have used the simplest explanation that suits the purpose at the time—provided it is not wildly inaccurate. Thus I have generally treated heat as a form of energy but have used the glossary to indicate that there are other interpretations.

A list of books and useful websites for further reading is also provided to enable the reader to take matters further if they wish. An appendix provides conversion factors between relevant units of measurement.

As a Brit writing for a largely American readership, I faced the interesting problem of differences in language. We tend to assume that all is well, because we can watch each other's films and read each other's books. As I have found in writing this book, it is not quite as simple as that. The Irish author George Bernard Shaw once said, "England and America are two countries separated by a common language." I now know exactly what he meant—particularly when it comes to technical language. Consequently, I have made extensive use of English-American dictionaries (such as http://english2american.com), and this book has been reviewed by a number of Americans (who, incidentally, sometimes couldn't agree among themselves). I take full responsibility for any remaining quirks of language.

CHAPTER 1

Feeling Good

The Science of Comfort

It's party time. To mark the midwinter solstice, you invite a small group of friends, say twenty, over to your house to celebrate. Very soon, even though it's close to freezing outside, the front door is ajar and the windows are wide open. Inside it has become uncomfortably hot. The reason? Human beings are little generators of heat energy, producing an average of about 100 watts at all times. Those twenty people at the party are therefore the equivalent of a 2 kilowatt fire burning in your living room. In fact, because this is an average figure, it is an underestimate. The energy produced depends on what you are doing—sitting around reduces the 100 watts to around 60 watts, whereas vigorous partying can raise it to 150 watts or more. The output of the party group is in fact much nearer to 3 kilowatts. No wonder the atmosphere gets heated!

Heating engineers are well aware of this effect and call this "extra" heat energy an *incidental gain*. All energy from household equipment ends up as heat energy, so another example might be your child's bedroom as she plays computer games while watching the television and listening to the stereo! Incidental gains can be a particular nuisance in a kitchen, for the heat from an oven can be considerable. As you can imagine, too many incidental gains can really screw up an engineer's carefully worked out figures.

Human biologists study the body far more closely than do heating engineers, and they are the specialists to whom we must turn for advice on how our bodies react to different conditions. An adverse reaction spells discomfort. Metabolic activity, the total rate of energy expenditure of humans, consists of two parts: the first is the energy we use to keep our lungs breathing, our heart pumping, and so forth; the second part is what we expend on our activities. To take the first part, for example, children with cystic fibrosis have their lungs clogged by viscous mucus. For them, breathing is a real chore, and they need far more energy to drive the air they need in and out of their lungs, which gives them a much higher metabolic rate. This, combined with their inability to absorb fats easily, means that children with cystic fibrosis usually are on high-fat, high-energy diets from the day they are diagnosed.

As for our activities, table 1.1 shows some metabolic rates for a typical young adult male. The metabolic rate is measured in watts per square meter (a square meter is a little over ten square feet) of body surface area. Sleeping and sitting don't burn too much energy, but brisk walking does. That's why, if you wish to lose weight, it's good to increase your exercise regime, since if you increase your metabolic rate some of the extra energy you need is eventually taken from your bodily store of fat. You can overdo things, of course, and your body has built-in mechanisms to avoid getting too hot from exercise. Roughly speaking, it's wise to avoid metabolic rates above 200 W/m^2 if you want to eliminate the nasty effects of overheating. The actual figures vary slightly from person to person, but for all of us, comfort is closely related to the metabolic rate.

Similarly, our comfort is strongly influenced by the surrounding air temperature, but my ideal comfort temperature may be 5°F higher than yours. This is the position in our household. I set our thermostats at a pleasantly warm 65°F. However, my wife finds this much too cold. So she turns it up to a sweltering 70°F. Early in our marriage, I used to point out how much money we would save by running the system at a more reasonable temperature. I soon learned that this approach makes her angry! Luckily, small "personal adjustments" to our clothing can easily resolve this matter. Discarding my tie and sweater helps, or I can take a shower and get into my summer clothes. Usually, however, I simply remind my wife of how much I love to see her in the beautiful

Table 1.1   Average Metabolic Rates for Young Adult Males

Activity	Metabolic rate (W/m² of body surface)
Sleeping and digesting	47
Lying quietly	53
Sitting	59
Standing	71
Brisk walking	154

sweater I bought her for her birthday, and ask why doesn't she put it on for me. This generally does the trick!

But why do layers of extra clothes work to keep us warm? Think for a moment of a polar bear. How can a warm-blooded mammal keep warm in the freezing temperatures of the Arctic? One important factor is its coat, which consists of very long hair. This hair traps air, and when air isn't allowed to move, warmth can only pass through it very slowly. Air is, in fact, one of the best insulators we know of. Putting on another layer of clothes traps another layer of air. We lose less heat. Our bodies therefore have to generate less heat, and our metabolic rate decreases, putting less stress on our systems. We feel comfortable again.

Our bodies can also respond differently to temperature changes according to the season. A cold snap in summer, when the temperature "plunges" by 15°F may feel very cold, even though it's still 60°F outside. But a warm spell in winter, when the temperature "soars" to the same 60°F feels extremely warm. The body has adapted during the intervening period.

Bodies have also evolved over hundreds of thousands of years. The short, squat Inuit peoples of Alaska, for example, have a smaller exposed surface area than the tall, thin Masai people of Kenya and Tanzania. A smaller surface area means a smaller heat energy loss, and is therefore advantageous in a cold climate. So, according to classical Darwinian evolutionary theory, smaller individuals are more likely to survive, and over time residents of a cold climate would become noticeably smaller. A further difference between the Inuit and the Masai people is that the Inuit have much greater fat-storage capacity to help them survive in the cold climate. As well as being an energy store, this fat forms a larger insulating layer directly under the skin, helping to keep the body warm.

In one episode of the sitcom *Seinfeld*, Jerry's curious friend Kramer falls asleep in his hot tub and, while Kramer naps, the heat pump freezes. He wakes up in 58°F water and complains that he can't get his core temperature up. During the rest of the episode, set in summer, he walks around in a long over-coat drinking hot tea. Normal human beings have core temperatures in the range of 98.4°F ± 0.8°F, as measured *sublingually* (with the thermometer under the tongue). The body has a number of defense mechanisms (called *thermoregulatory mechanisms*) that help it maintain this core temperature, which is the optimum temperature for "normal" biochemical processes to run efficiently in a healthy person. Sweating and shivering are two very obvious examples of these thermoregulatory mechanisms. When we have an infection our bodies often respond by raising the tissue temperature to threaten the invading bacteria. The downside of this is that we feel ill, which forces us to rest awhile so our bodies can try to recover. Even a slight rise of 1°F can produce this response.

I well recall my 4-year-old son waking one night and crying with the pain of earache. My wife and I soon realized that his temperature had risen too, and we hit the panic button. Once the doctor had been sent for, we set about trying to lower his temperature, using damp cloths, wet towels, and so forth. We now know that this is not necessarily the best response, because a fever may not be a bad thing. Mind you, once the temperature is more than 5°F above normal, it really is a panic situation. It depends on just how severe the patient's fever is. Since realizing this, we have always carefully taken the temperature and sought over-the-phone advice from the doctor before doing absolutely anything. Fortunately, my son soon recovered with the right antibiotics.

Warming Up

If you're living in a temperate climate, chances are your body is hotter than the environment, which means that heat energy moves from your inside outward to your skin (the process is called *conduction*). The outcome is a temperature gradient from your inside to your outer skin, with your skin having a lower temperature than your inside. The air near your body also has an insulating effect. There is therefore a further temperature gradient arising from the difference between your skin temperature and that of the surrounding air. The outcome is a boundary layer of air, an inch or so thick, that acts like a thin layer of clothing. The loss of this layer in a breeze leaves your skin directly exposed to the surrounding air—like suddenly loosening your clothes—and you feel much colder, even though the air temperature has not changed. The air has merely moved. This is known to meteorologists as the *wind chill factor*. Comfort, therefore, is not just a matter of having the right surrounding temperature. But more of this later.

When the surrounding air heats up or cools down, your body reacts accordingly. When it gets cold, your brain, specifically the hypothalamus, sends a signal to your body, and you start to shiver or produce goose bumps. Shivering forces your body to move, and because it is moving, it generates more energy. So, shivering warms you up. In recent medical studies, patients who had undergone heart attacks were cooled down while surgery was performed. The idea is that restoring heart function can cause a surge of blood to flow into the brain, causing inflammation, bruising, or brain damage. Keeping the patient cold helps prevent this. The main difficulty, doctors have discovered, is that the patient has to have medication to prevent them from shivering. The body's innate attempt to keep warm means it presents a moving target to the surgeon.

Goose bumps, on the other hand, are leftover from the days when we, like the polar bear, had a long hairy coat—hairs are more effective at trapping air next to our skin when they are standing on end.

If you're very cold, your natural response might be to curl up into a ball—the fetal position. An interesting property of a sphere is that, of all geometrical shapes, it has the least surface area for any fixed volume. As heat loss depends on surface area, by curling up you reduce your surface area and stay warmer for longer. Of course, you could try putting on more clothes instead. My wife really does look beautiful in her new sweater!

Stricken with frostbite during Captain Robert Falcon Scott's unsuccessful attempt to get to the South Pole, British hero Captain Lawrence Oates uttered the memorable words, "I'm going outside; I might be some time," choosing to go to his death in the vain hope that it would conserve supplies for those who remained. But what is frostbite? Faced with extreme cold, the body reduces the amount of blood flowing to the extremities—toes, fingers, nose, and ears—as it tries to keep the core temperature high. The first consequence is that the person looks pale, as less blood is reaching the skin. Prolonged shortage of blood flow to the extremities can lead to the skin and underlying tissue freezing and dying. This is frostbite.

A condition related to frostbite, hypothermia, has symptoms that include "the umbles"—stumbles, mumbles, fumbles, and grumbles—which are surefire signs of the loss of higher motor function in the brain brought about by the body trying to conserve its heat energy. It is a serious condition that requires careful medical attention. Anyone with a core temperature of less than 95°F is classed as hypothermic—the clinical thermometer in your home first-aid kit isn't just for checking up on fevers.

Cooling Down

Once your body starts to get a tad too warm, the blood vessels dilate and more of your blood is pumped to the far reaches of your skin, from where the heat can more easily escape to the surrounding air, leaving your body cooler. As a result, overheated people look red. This reaction is accompanied by sweating. Droplets of water form on your skin and heat flows from your skin into them, producing evaporation and cooling. In a humid climate—Disney World in July, for example—it's much harder to get the water into the already-saturated air, leaving you still hot but also clammy. There's a price to be paid for producing sweat though: you start to become dehydrated, for you can easily lose as much as two pints per hour on a hot day. Those who don't drink water on hot days run the risk of heat exhaustion or heat stroke.

As a middle- to long-distance cross-country runner in my youth, I learned about heat stroke the hard way, when I collapsed after five miles of a six-mile race on a hot spring day. Scary, but no great harm done. Nowadays there are refreshment stations at convenient places along the course, and a whole new set of products called isotonic drinks has become available to athletes. These don't just replace lost fluid but lost body salts as well, and have the optimum concentration for ease of bodily absorption. A low salt level in your muscles leads to painful cramps, which can be an early symptom of heat exhaustion and a signal to rehydrate your body before your symptoms worsen.

You can also use a fan to cool off, which drives cool air past your skin. And, as beachgoers know, you can take more clothes off—though this may increase your risk of getting skin cancer. Baggy clothes also help—it is no accident that the national dress of many Arab countries consists of loose-fitting, lightweight garments. These work in two ways; first, they protect the skin from the direct heat of the sun, much of which is either absorbed or reflected by the cloth. They also catch the slightest breeze in the air and effectively create a breeze across the skin beneath the garment as the wearer moves along. This time, loss of that boundary layer turns out to be useful.

Building for the Body

The human body can cope with a tremendous range of ambient temperatures, roughly 80°F ± 20°F, without experiencing major discomfort. We can control our temperatures by adding or subtracting clothing or through sweating, shivering, and other biological means. All of these, however, put some stress on our bodies, and as a result we, to varying degrees, feel uncomfortable. A good rule of thumb to remember is that an average indoor winter temperature of 68°—adjustable by about 7° in either direction as needed—will suit most people.

We can design buildings that achieve this level of control by installing a heating system to keep us warm. In *Little House on the Prairie*, a series of books by Laura Ingalls Wilder made popular in the seventies and eighties by a television series, a pioneer family in North Dakota manage to survive a harsh winter by burning knotted straw in an old cooking stove. Modern heating is more sophisticated: forced-air systems can blow hot or cool air through the house, and we can regulate the temperature in the room by controllable vents.

Many of these systems, however, rely on fossil fuels (coal, oil, or gas), and there is a limit to the amount available over the long term. How long we can continue pumping the waste gases from these fuels into the atmosphere before we significantly and adversely affect our climate is also a major concern. In some places, solar heating can help reduce bills and overcome environmental concerns by allowing the sun to do its part in keeping things warm. We'll look at such systems more closely in a later chapter.

Boiling Hot

Designing a building for year-round comfort is not as easy as we might think; it involves far more than protecting us from an adverse climate—whether that be the cold Arctic or the hot tropical plains of Africa. With good design we can create an indoor "climate" closely matched to the comfort needs of our bodies wherever we are. This is achieved largely by controlling the processes of conduction, convection, and radiation.

Take a saucepan full of water and put it on the stove. It will, eventually, start to boil. What's happening is that the heat energy from the stove results in "packets" of water moving about within the saucepan, an example of *convection*. That's markedly different from *conduction*, where heating something from below causes the layer above it to be heated, which in turn heats the layer above that, and so forth, and in which nothing moves at all.

Convection is brought about when a liquid or gas is heated, making it less dense and thus more buoyant. A blob of warm liquid will then sail upward through the surrounding colder liquid. Just think for a moment: you put the kettle on to brew a cup of tea, and when it is boiling, you pour the hot water onto the tea in the bottom of the pot. If conduction were the only heating mechanism, then the result would be a strong brown sludge on the bottom of the teapot. Without convection, it would be impossible to brew a decent cup of tea, and civilization as we know it could not have developed. (Okay, your author is a slightly eccentric Brit, prone to a little exaggeration now and then.)

Exactly the same thing happens with gases—most importantly, air. Wood smoke from a campfire can travel by convection to quite a considerable height on a still day—a fact not lost on Native Americans, some of whom used this process for signaling over long distances. As our skin is usually hotter than the surrounding air, we are cloaked in microcurrents of moving air almost all the time—a useful mechanism for cooling our bodies. These are natural air currents, affecting the boundary layer referred to earlier, but their effect can be augmented by artificial means.

Convection, consequently, creates heat losses that depend on a few things: temperature difference, surface area, and the rate of movement of the surrounding air. The rate of heat transfer increases as the temperature difference between your skin and the surrounding air increases. A high body temperature therefore raises this difference, and with it the rate of loss. The hotter you are, the more rapidly you will be cooled. Larger people, that is, those with a large surface area, also cool more quickly than smaller ones—remember the Inuit and the Masai mentioned earlier?

Have you ever stood near a cooling fan? You feel cooler if the fan rotates more rapidly, as this moves the air over your skin more quickly. This effect is the same as the wind chill factor discussed previously. Nature can achieve the same thing via the wind: compare two days for which the air temperature is the same 68°F, but on one of the days there's a stiff breeze blowing rather than still air. You will feel cooler on the one windy day. According to the National Weather Service, a 10 mph wind on a 32°F day results in a temperature that feels like 14°F. To estimate the wind chill factor, meteorologists use models incorporating the wind speed at a height of five feet (the height of the average human face above the ground), skin-tissue resistance, and the thermodynamics of the boundary layer of air surrounding the body.

I recently told one of my students that I still occasionally looked at my boyhood stamp collection. I was teasingly told that I was an "anorak"—a word that now means someone who is obsessed by trivia. When younger, I used to go walking in the hills wearing the latest available protective clothing. It was a double-layered, closely woven, cotton covering that was worn like a sweater. The cuffs were elasticized, and there were drawstrings around the waist and hood. All this to keep the wind out! This garment was called an anorak. The new meaning came about because another group of people habitually wore anoraks—trainspotters who hung around draughty railroad stations writing down the numbers of the trains they saw. Outdoor garments have improved since those days—but wind chill will always be a major consideration in their design.

Most people prefer, and feel more comfortable in, a room in which the air circulates. For a sedentary person, air speeds in the range of four to six inches per second are acceptable. Any less and the room feels stuffy; any more and there's the uncomfortable draft that my great-aunt Hilda always complained of. If you're working hard, perhaps in a factory doing strenuous labor, an air current of twenty inches per second is more like it. This simply reflects the fact that the worker has a higher metabolic rate than Great-aunt Hilda so needs a greater cooling rate to avoid the effects of heat stress.

Cold Feet

As most people know, hot air rises. This is another example of convection and means that the basement of a house is often far cooler than the attic. In a big house you might need to have two-zone heating, so that the temperature control can be set differently upstairs and down. In a poorly designed building, there can even be a noticeable temperature gradient between ceiling and floor. A tall person might have a hot head but cold feet.

My first experience of factory work was when I took a part-time job in a detergent factory to supplement my income as a student. My job was to fill gallon cans with liquid detergent and fasten the top. Relentlessly, the cans kept on coming down the line, and it wasn't long before I was bored out of my mind. Yet I was physically comfortable. I was seated for most of the day and therefore close to the floor (and so immersed in cool air). Our supervisors, who kept watch from a gallery above the shop floor, complained of the heat. At first I could never figure out why—but it was of course the hot air, which had risen to above ground level. We all would have felt far better if the management had invested in a decent heating, air-conditioning, and ventilation system!

Nowadays a *destratification fan* could be fitted in the ceiling. These rotate slowly and push the warm air back toward the floor. With careful adjustment of the HVAC (*heating, ventilating,* and *air-conditioning*) system, they save more energy than they expend and are therefore worth the investment. These fans can be noisy, though. We had one installed in our local church. Some members of the congregation find the slight hum it makes distracting during the quieter parts of the service.

Completely Fried

The sun shines and the earth gets hot. There's no conduction, nor any gas or liquid to allow convection in interplanetary space. The heat energy is transported from sun to Earth by another process—*radiation*. The amount of heat radiated by a surface depends on its temperature and the properties of the surface. The high temperature of the sun allows enough radiated heat to reach the earth to sustain life.

You don't need huge temperatures to notice the effect of radiation. In temperate and cold climates, our bodies radiate heat outward. Fortunately, we also absorb radiation from objects around us. If the radiated and absorbed heat is balanced, we stay at a nice, stable temperature. Interestingly, merely standing next to a cold surface can make you feel cold, because you radiate more heat outward than you receive from the cold surface. Try standing next to a cold, closed window and then compare how you feel standing next to a warmer adjacent wall. Air temperature is therefore not the only temperature that matters—nearby surfaces within the room are also important. Similarly, try (very carefully!) putting your open hand within an inch or so of a hot iron. You will feel the radiant heat as it warms your hand.

On a warm day, by the same principle, the outer brickwork of a house can get really hot. At night, when the air temperature has dropped, the bricks, like the iron, will radiate their stored heat back out into the nighttime air. The same idea is at play in heat storage systems such as electrical underfloor heating or thermal storage radiators.

Astronomers look at asteroids and try to work out good theoretical models by which to describe them. One thing they are concerned with is the *albedo* of an asteroid. This is the proportion of light and heat from the sun that the asteroid reflects back into space. The concept of the albedo can be useful for thinking about heating our homes as well. White walls have a high albedo, reflecting back most of the radiation that strikes them, while black walls absorb it. So, in a hot climate, paint your outside walls white to reduce radiative heating of the house. It's also better to wear white clothes in hot climates so that the radiation is reflected back into the air, rather than absorbed, which would heat you up.

How comfortable we feel within our home is not only a matter of the air temperature inside it but also of the amount of radiation our bodies receive. Because our skin absorbs much of the radiation that falls upon it, radiation can be an efficient and economical form of heating. It should be no surprise therefore to learn that household radiators have been designed to give out a large proportion of their heat by radiation. Scientists have studied our body's reaction to radiant heat in some detail and have developed the concept of *mean radiant temperature,* or MRT, to represent the amount of radiation received. It is calculated by a rather complex formula from the temperature of surrounding surfaces and their orientation, and it is this acting together with the air temperature that determines our comfort.

Table 1.2   Four Conditions of MRT and Air Temperature
That Give the Same Comfort Levels

Mean radiant temperature (°F)	Air temperature (°F)
65	75
66	71.5
67.5	67.5
70	65

Each of the four sets of conditions shown in table 1.2 gives close to the same feeling of comfort, assuming no change in air movement. For example, we feel as comfortable when the MRT is 65°F and the air temperature is 75°F as when the MRT is 70°F and the air temperature is 65°F. Therefore, an increase in radiant heat can make up for a decrease in air temperature. So you might feel okay simply by turning up any appliance that increases the radiant heat—an electric bar fire for example. The table also shows the importance of warm walls. A drop in air temperature due to someone opening a door and letting a howling gale blow in momentarily won't feel so bad, because the radiation from the warm walls will not have changed.

The igloo is an extreme example of how a variety of factors operate to produce a building's internal temperature. Even though the MRT is low (below freezing), and the internal heating is simply that produced from the bodies of people inside it (98.4°F), it is a successful building for its climate. There is little air movement inside, and therefore no wind chill, low convective losses, and excellent thermal insulation. The high albedo of the internal walls, which are white, also means that any radiant heat trying to reach the outside is simply reflected back in. The result is a stable internal temperature above that outside. Moreover, an igloo can be quickly built from plentiful local materials.

Sticky Stuff

Other things affect our comfort. Conduction can come into play if you like to walk around the house barefoot. On a cold day, bare hardwood floors are chilly, but a carpet feels reasonably warm. This has to do with the thermal conductivity of the materials, a measure of how easily heat energy is carried away by a particular material. Your feet rapidly lose energy to the concrete or wood floor, but the low thermal conductivity of carpet makes it an effective insulator, which helps retain the heat in your body.

A leather couch poses a different problem. It can be refreshingly cool to sit upon on a warm day, but if the sitter sweats, they can be left stuck to the surface. Of the heat energy our bodies pump out into the environment, a surprisingly high 10% comes from evaporation. The water vapor comes from "externally exposed surfaces," such as your skin, the inside of your mouth, and the inside of your nose. If your nose gets blocked because of a cold, you have to open your mouth more to breathe, which results in far more moisture evaporating from that region than usual. The result is a dry mouth and a sore throat. The leather couch on which you sit doesn't let the water evaporate into the open air. It just stays there, drenching your shirt and covering the leather with a thin film. Luckily, leather is easy to wipe clean!

Sweating is a good way to keep cool when it's hot, especially in a dry climate. In muggy conditions, it doesn't work so well. Dehumidifiers can help by extracting the moisture from air. Not only can they keep basements dry and mold-free, the lower humidity makes sweating more efficient.

Humidity, scientifically speaking, is either absolute or relative. Absolute humidity reports the amount of water in a fixed volume of air (such as the number of ounces of water in a cubic foot of air). Relative humidity is the absolute humidity expressed as the percentage of water needed to saturate the air. So, if the air is completely saturated with water vapor, the relative humidity is 100%. Perfectly dry air would have a relative humidity of 0%. The comfort range covers a relative humidity between 40% and 70%. Figure 1.1 combines average comfortable humidities and temperatures to define a comfort zone for sedentary activities.

The other side of the humidity coin has to do with condensation. Suppose the air contains a certain amount of water per cubic foot. Then night comes and the air cools. Cold air has less capacity to hold water vapor than does warm air. The result? Condensation. The temperature at which condensation starts is known as the dew point, for the dew that you can see on the early morning meadow is simply water that has condensed out of the cold nighttime air.

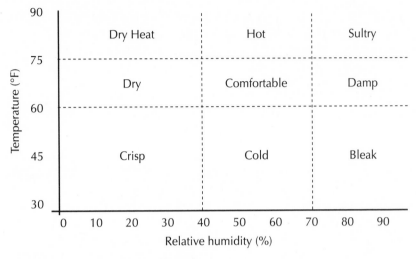

"Sultry" is oppressive heat and humidity.
"Bleak" is unpleasantly cold and damp.
"Crisp" is cold and invigorating.

Figure 1.1.   The combined effects of temperature and humidity upon comfort

Indoors, condensation is, for example, the mist on the bathroom mirror when you take a long shower. Some of the droplets from the shower go into the atmosphere and then condense again as droplets on any handy cold surface. The bathroom mirror is one such surface. When I was a young boy I often went to bed in the depths of winter in an unheated room and woke up to find my bedroom window frozen over to form beautiful patterns of ice. Mom would say Jack Frost had been there that night! What had actually happened was that each time I breathed out, I put more dampness into the air. Ventilation was kept to a minimum to reduce drafts, so the damp air couldn't escape. The result was an increase in humidity, which condensed on the cold surface of the window, where it slowly froze. Jolly uncomfortable, but beautiful nonetheless.

While condensation doesn't pose too many immediate problems for humans—we can still carry on singing in the shower even when we can't see ourselves in the mirror—the long-term effects can be unpleasant. It can ruin your favorite *objet d'art* for a start, as well as providing ideal areas for mold to grow. Some people have an allergy to mold, and those with compromised lung function can sometimes end up with mold, such as aspergillus, growing in clumps in their airways. Improving the housing stock of a nation can improve the overall health of its citizens by reducing condensation. Dehumidifiers and improved ventilation are two ways of doing this. Simply increasing the internal air temperature by improving insulation or installing a better heating system will also do the trick.

Sitting Comfortably?

I have happy memories of trying to help my sons with their physics homework. The problem, however, was that they kept asking questions and demanding further explanation. But when I explained why, they complained bitterly that I was making it too complicated and asked whether I could please just give them the answer!

Science is like that—you start with something that seems quite simple and soon end up enmeshed with factors and variables you had never thought of. Nonetheless, I hope you have enjoyed the scientific explanations in this short account of comfort. However, for those of you who just want the answer, I've summarized comfort conditions in table 1.3. I hope you find it useful.

The factors shown in the first column of the table are uncontroversial, and similar lists are common elsewhere. The second column gives broad guidance, and figures vary slightly from author to author. This is largely because different authors specify different conditions in arriving at their recommendations. The third column indicates some of the control measures available to achieve comfort. It is worth noting here how the same control measures are able to influence a number of factors. Most of these are covered in subsequent chapters of this book: insulation and ventilation in Chapters 2 and 3, heating systems and their effectiveness in Chapters 5 and 6, and temperature control in Chapter 7. I also hope that you find the tips in box 1.1 useful.

Table 1.3   Factors Affecting Your Comfort

Factor	Comfort conditions	Control measures
Metabolic activity	Rates below 200 W/m²	Food intake Voluntary control of activity Level of clothing
Air temperature	60°F–75°F. Internal design temperature for heating is taken as 68°F	Metabolic activity as above Effective heating system Good insulation
Conductive losses	Little direct contact with cold surfaces	Clothing (gloves, socks, shoes, etc.)
Air movement	4–6 inches/sec (sedentary)	Controlled ventilation
Temperature gradient	Less than 3°F difference between temperatures at floor and head levels	Insulation Destratification fans
Radiation	Mean radiant temperature close to the air temperature	Warm surfaces (good insulation) Effective heating system
Moisture	40%–70% relative humidity Wall temperatures above dew point	Ventilation and temperature control Insulation and temperature control

Box 1.1. Top Tips for Saving Energy while Maintaining Comfort

1. Use incidental gains: Turn down your fires and heaters rather than shed clothes when the room gets too hot.

2. Make small personal adjustments: If a room is too cold for you, but just right for your partner, then add clothes.

3. Dress according to conditions: loose, light-colored, lightweight garments for hot weather; several layers of thicker clothes for cold weather.

4. Control ventilation: too much wastes energy, and both too much and too little are uncomfortable.

5. Insulate your home to save energy and money. Comfort will also be increased.

6. Control humidity: Install well-designed and carefully placed air-conditioning units, though either humidifiers or dehumidifiers may be a cheaper and effective solution in some situations. They would also use less energy than air-conditioning units.

7. Control temperature: Set your thermostat(s) as low as is comfortable in the winter and as high as is comfortable in summer.

CHAPTER 2

Save a Dollar, Save the Earth

Effective Insulation

Free Lunches!

The US economist Milton Friedman once said, "There's no such thing as a free lunch." This is a quotation with which I am very familiar, for I have lost count of the number of times my students have quoted it to me. They know of my enthusiasm for energy and are of course enjoying winding me up. Yet I really believe that insulation can provide any number of free lunches, once a relatively small initial investment has been made. As a bonus, it considerably increases comfort too.

For example, have you ever suffered from the chilly drafts that come under the doors on a cold windy night? A simple solution is to put a long piece of material in front of the cracks that cause the problem, thereby blocking the movement of the incoming cold air. This can not only stop cold air coming in but also stop warm air going out, so less fuel (and money) is needed to heat the air in the building. The cost should easily be recovered within a year, and thereafter the savings can pay for lots of free lunches.

Insulating the building can produce even more savings. How come? In a large field near my boyhood home lie the ruins of a medieval castle. The walls are several feet thick to prevent successful attacks. I used to imagine the various siege engines that might have been used to breach the castle walls and wondered how the occupants kept warm in winter. The thick walls were the key: the thicker the walls the slower the energy loss, and relatively little wood would have to be burned to replace it. The importance of this is obvious when you think about a castle under siege—you cannot go out to get more firewood! A high level of insulation leads to less fuel used over the period of the siege. The same applies to our homes. The lowest-cost option for obtaining a comfortable internal temperature over a winter is usually to increase the insulation.

The benefits of insulation, of which draft-proofing and thick walls are examples, go far beyond money saving. Remember in Chapter 1, the nighttime visits of Jack Frost to my bedroom windows? Double-glazing sends Jack away, because the glass is a bit warmer, thereby reducing condensation. The same happens with exterior walls where condensation not only ruins decorations but can also lead to fungal growth and the release of spores into the air. Jeffrey C. May, in his book *My House Is Killing Me!* tells of a house in Texas that became so contaminated with a toxic mold that the family had to abandon the property and bulldoze the house. Adequate insulation would have prevented the original condensation, which was the root cause of the problem.

The increased wall temperature has another positive effect. Have you ever stood close to an exterior wall, inside your house, on a cold day? Your body radiates heat toward the cold wall and you feel cold even though the internal air temperature is okay. Insulation leads to a warmer wall, and so you radiate less energy, feel warmer, and can turn down the heat a bit more.

My son Andrew and his wife, Ann, recently came to stay awhile. Ann was brought up in the industrial city of Preston in the UK. Staying in the countryside is a great pleasure for her—except for her sleep. The baaing of the sheep and hooting of the owls keep her awake. It's not just noisy roads, aircraft, or factories that cause noise pollution! The good news is that thermal insulation has a beneficial acoustic effect as well. If you wrap your head in a towel, you cannot hear as well. Insulating a building has the same effect—less noise penetrates inside.

But you are not the greatest beneficiary of your investment. A well-insulated home helps save the planet. When oil, coal, and natural gas are burned, they produce carbon dioxide, which has been implicated in global warming—and

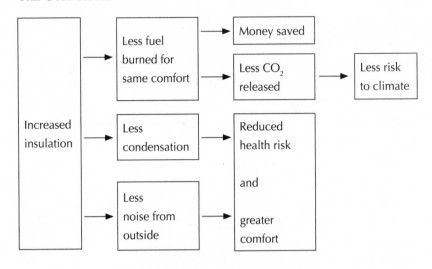

FIGURE 2.1.   Benefits of insulation

thus in potentially catastrophic climate change. Yet "nothing will come of nothing," as King Lear said, for carbon dioxide is released in the manufacturing and installing of the insulation. Does this wipe out the benefits of insulation? A detailed look yields a resounding "no" to this question, largely because insulation lasts the life of the building and needs no maintenance. Over the long term the benefits outweigh the initial costs.

Although this (and the next) chapter is written largely about the need to keep warm, it may surprise you to know that exactly the same principles apply if the aim is to keep cool. The ancient Egyptians retired to the cool of underground chambers on hot days, using the earth as a very effective insulator. Similarly, the early Spanish mission houses in the southwestern United States were relatively cool due to their thick straw and clay walls. Nowadays, in a hot climate, good insulation saves money by reducing the time for which the cooling system or air-conditioning operates. In fact, since the efficiency of the heating process is greater than that of air-conditioning, it is more important to have well-designed insulation in a hot climate than a cold one—the savings in both energy and money can be much greater.

The range of benefits of insulation are summarized in figure 2.1. It saves money, increases comfort, *and* helps save the planet. A win, win, win situation!

Bliss—A Warm Steady Temperature

As I lie relaxing in the bathtub after a day's work, I occasionally dislodge the plug, causing the water to start leaking away. Imagine for a moment that I cannot get the plug to fit back properly. If I turn on the taps, the tub starts to fill again, and if I carefully adjust them until the amount seeping out equals that dribbling in, the water level stabilizes. Achieving a steady temperature in a building is similar. Once the rate at which heat energy enters equals that at which it seeps out, the temperature stabilizes. The trick is to stabilize it at a blissfully warm, comfortable temperature.

Can I simply turn the heat up in my house to warm it up, and then turn it down when I feel too warm? Yes I can, but using a thermostat does this much more easily for me. Full details of thermostats are given in Chapter 7, but basically they bring in lots of heat energy until the building is at the required temperature, and then control the heat coming in to match that going out, leaving us with a warm stable temperature. This is most economically done when there is the lowest possible amount of heat seeping out, for then only a slow input is needed to match it. The purpose of insulation is to slow the out-flow of heat so that only a slow input is needed to replace it.

In practice, there are two ways in which heat energy can be lost from a house (like having two plugs in your bathtub), and both should be minimized. That which flows through the walls, roof, windows, and floor is known as the *fabric loss* (i.e., the loss through the fabric of the building). In the old days, if a house caught fire, the village folk would form a line leading to the village pond and pass buckets of water from one to another. Much fabric loss occurs by conduction—energy passing from molecule to molecule from the warm inside surface of the wall to the cold outer surface, where it disperses to the surroundings—a bit like the water passing down the line to be thrown onto the fire.

The other loss is from ventilation—the clean fresh air infiltrating the building. This is essential if you are not to die through lack of oxygen, but whenever cold air comes in, warm air also goes out. This is the *ventilation loss*. The draft excluder behind the door is one way of controlling it. Ideally you need enough ventilation for your health, but not so much that you feel it in your pocket through high fuel bills!

But, before we look in detail at the myriad ways of controlling these losses, we need to think about insulating materials. I've drunk hot tea from both a Styrofoam cup and from an aluminum one. Believe me, there's a difference in how they perform thermally!

Resisting the Flow

Back in the mid-eighties my then boss (a bit of a control freak), attended a management course. She came back with that well-known mantra of management, "If you cannot measure it, you cannot manage it." This was mentioned at every meeting and on every document coming out of her office. Many hours were spent identifying important parameters and devising "indicators." To many of her employees it was a huge turnoff, and morale plummeted. She never knew this though—for she failed to develop a "morale indicator!"

Yet in terms of physical processes, like heat loss, the slogan is exactly right. We need an "indicator" to measure the effectiveness of insulation. The one most commonly used in the insulation industry is the *thermal resistance* (or *R-value*), which is a measure of how much the material resists the heat flow. Technically it is:

the resistance to heat flow through unit area of the material caused by unit temperature difference across the material.

The higher the R-value the better the insulation and the lower the rate of heat loss.

Yet the area through which the heat escapes is not the only issue. The epic adventure novel *Moby Dick*, by Herman Melville, published in 1851, rests on the commercial value of the whale blubber from which oil was made. Blubber forms an insulating layer around the whale's body, which retains heat even when surrounded by the cold sea. It can be eighteen inches or more thick, increasing in the winter months. In fact, some large whales, swimming in Arctic or Antarctic waters, may have about 70% of their body mass made up of blubber. Which starts me wondering. I generally put on weight in the winter too. Perhaps this is a normal reaction to lower air temperatures and not due to insufficient exercise or overeating at Christmas after all!

Table 2.1   R-values of Some Common Insulating Materials

Insulation material	R-value (ft² °F hr/BTU per inch)
Loose fill	
Fiberglass	2.2–2.9
Rock wool	2.2–2.9
Vermiculite	2.2
Perlite	2.7
Cellulose	3.1–3.7
Blanket material	
Fiberglass	2.9–4.2
Sprayed insulation	
Polyurethane foam	5.6–6.2
Icynene	3.6–4.0
Board material	
Polystyrene (extruded)	5.6–7.0
Polyurethane	4.4
Polyisocyanurate (unfaced)	5.8–6.2
Polyisocyanurate (faced)	7.1–8.7

Sources: www.energyguide.com and www.hometips.com/cs-protected/guides/insulation.html

So, the insulating effect depends also upon the thickness of the insulating material. For this reason, the R-values are usually quoted "per inch," since the effect generally increases linearly with thickness. Thus, a two-inch thickness of a material will have twice the R-value as one inch. Table 2.1 compares some common insulating materials. Since the value can be affected by factors such as temperature and density, a range of values is quoted. For more accurate work, R-values have been tabulated by ASHRAE (the American Society of Heating, Refrigerating and Air-conditioning Engineers) and can be looked up in their handbook. The Federal Trade Commission also finds this information useful, for it is responsible for controlling the testing of insulation products. It protects customers from misleading claims by mandating that R-value information be disclosed in ads and at points of sale.

But R-value is not everything. There must be mechanical strength too. One of the joys of parenthood and grandparenthood is that you get to read again all the well-loved stories you knew as a child. One such story is that of the three little pigs and the big bad wolf. The first pig built his house of straw, the second of sticks, and the third of bricks. Naturally only the brick house was able to stand the "huffing and puffing" of the wolf, who blew the other houses down. Building houses out of insulating materials would be rather like building them from straw: lovely and warm but with no real strength.

Figure 2.2 shows a cross section of a typical timber-framed wall. It is a composite structure firmly fixed to a strong rigid frame, which gives it its strength. There are five layers, in front of which there has also to be siding (not shown), making six in all. The R-value of the whole structure is the sum of the R-values of the six individual layers (strictly it is eight layers, for the narrow layers of air, an inch or so thick, in contact with the inner and outer surfaces should also be included, though their contribution is relatively small).

This discussion of R-values also assumes that there are no gaps around the finished wall through which cold air can creep in or warm air sneak out. Leaky insulation can happen if the caulking around the windows or doors is inadequate. So this needs to be checked from time to time. An even worse scenario in terms of heat loss occurs if you habitually leave doors and windows open in cold weather. The ventilation losses can then be enormous.

As well as for architects and heating engineers, the concept of R-value is very useful for governments. Many people both in the US and the UK feel that we are overgoverned, but perhaps they should remember the approach of John Locke, the seventeenth century English philosopher, who said in his *Second Treatise on Civil Government* that "government has no other end but the preservation of property." It may surprise you to learn that the US government finds the concept of R-value useful for precisely the reason given by Locke: to preserve property. State and local building codes set minimum standards for the different parts of new buildings. These standards specify minimum R-values. As well as reducing condensation and dampness, bringing all homes up to these values would be a good start in reducing the nation's energy bills. NAIMA, the North American Insulation Manufacturers Association, estimates that 60% of American homes are underinsulated. Upgrading would put many

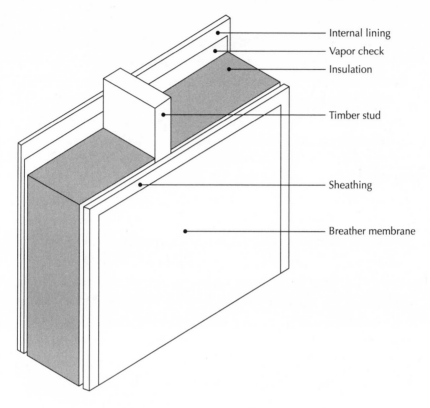

Internal lining
Vapor check
Insulation

Timber stud

Sheathing

Breather membrane

FIGURE 2.2.   Structure of a typical timber-framed wall, siding omitted

people in the free lunch situation—the cost of the insulation would soon be
recouped in lower energy bills.

Aren't statistics annoying? It's all very well to quote a figure of 60%, but what
you and I really want to know is how well we are doing and what else we need
to do. One way is to compare our level of insulation with that recommended by
the US Department of Energy. In their booklet, "Energy Savers: Tips on Saving Energy and Money at Home" (downloadable from their energy efficiency
and renewable energy website www.eere.energy.gov/consumer/tips/pdfs/energy
_savers.pdf), there is a list of recommended R-values for homes, broken down
by region. Alternatively there is a handy calculator, which just needs your zip
code, at the website of the Oak Ridge national laboratory, www.ornl.gov/sci/

roofs+walls/insulation/ins_16.html. These show that in most cases there should be at least R-30 in the attic, R-19 in the side walls, and R-19 for the floor.

How the recommended R-values are achieved is up to you and your contractor. For example, you can increase the R-values of masonry walls by filling the cavity between the inner and outer leaves with insulation or by using special insulation blocks for the inner leaf. But before you rush off to find a contractor, there is a further important consideration—keeping it dry.

Staying Dry

One of the corollaries to Murphy's Law (which incidentally, we Brits know as Sod's Law) applies here, namely, "things are never as simple as they first appear." The problem is that insulation's effectiveness is apt to decrease if it gets wet.

One of the lovable eccentricities of the English is that we will happily play our national game of soccer in pouring rain in winter, yet stop playing cricket (our other national game) at the first drop of rain in summer. When I played soccer I was the goalie, which often entailed standing around while play was at the other end of the field. Many times I got soaking wet and bitter cold, for goalies in those days wore a thick woolly sweater. These trap air in their fibers, keeping the wearer warm as long as they are dry. When the sweater gets wet, the water, a better conductor of heat energy than air, replaces the air and heat escapes from the body. If it stops raining, the situation can worsen, for the water may start to evaporate, leading to even more cooling—just like sweating.

Moisture is constantly being generated in a home. Around five pounds of water vapor is generated per day from our appliances (such as dishwashers), and additional vapor is released when we breathe or take a shower. Furthermore, just as heat energy moves from where there is a high temperature to where there is a low one, so water vapor moves from regions of high humidity to regions of low humidity.

The outcome is that water vapor will migrate through many building materials. This may seem strange, but remember that even the tiniest pore (in gypsum plasterboard, for example) seems like a massive cave to molecules of water vapor. So they may well wind up inside the insulation, where it is colder than inside the house. There they can condense (like the mist on your bathroom mirror) and the insulation may get wet. This is likely to happen to fibrous (as well as porous) insulating materials such as fiberglass, but is less of a problem for other insulating materials such as polystyrene and polyisocyanurate, which have different surface characteristics.

Look again at figure 2.2. You will see that there is a *vapor check* (also called a *vapor retarder* or a *vapor barrier*) between the internal lining and the insulation. This is often a polythene sheet, which greatly reduces the rate of moisture transfer. To be fully effective, however, any gaps, such as those around electrical fittings, must also be sealed. If you live in a warm, humid climate, such as around the Gulf of Mexico, there is often greater humidity outside the house than inside. The flow of water vapor is therefore reversed. In this situation the vapor check is placed on the outside of the building. I told you it wasn't simple!

In fact, Murphy must be having a good laugh at our expense, for it gets worse. During my time in East Africa, I lived for a short while in a prefabricated house. This was put up in the dry season and was perfectly serviceable until we had our first storm of the rainy season. Water then poured in through the roof, which was made of corrugated metal sheeting. I didn't know what to move first into the few dry bits of the house. I was torn between saving my teaching notes and trying to protect my bed so I could at least get a good night's sleep. Well, what would you have done in my situation? (I saved my notes on the basis that they would take longer to replace.)

The rather obvious fact is that the entire outside of a house needs to be impervious to water. Yet, putting a vapor check on the outside of a wall creates as many problems as it solves, for water vapor can be trapped within it. This needn't be rainwater. It could simply be some leftover from the time of construction. Gypsum finishing plaster is put on wet, for example, and any porous material that has been stacked outside will retain some water within it. Timber, in particular, can absorb significant amounts. Anyone who has moved into a newly built home will know that drying out can take some time—up to a year is not unusual, depending on the climate. The solution is to fit *breather membranes* (also known as air barriers, air retarders, or simply housewrap), as shown in figure 2.2. These allow water vapor to diffuse easily outward through them but are impervious to liquid moving in. As an added bonus, they act as a wind barrier to restrict the flow of air through the wall.

This is fine for walls, but what of attics? If sufficient ventilation exists, then condensation problems do not occur in attics in *most* US climates. But what then is sufficient? Guidelines are available from a number of sources (for example, www.betterinsulation.com/AtticInsulation.pdf), but these need to be verified with local practice and building codes. The situation with floors is similar.

Overall, keeping the insulation dry is a complex business for which there are no truly reliable rules of thumb. It is nonetheless important to be aware of potential moisture problems, if only to prevent ourselves making matters worse when we set about "improving" our homes.

More and More Means Less and Less?

My formative years were the sixties, when flower power was rampant, protest demonstrations were common, and more than a few folk were on mind-bending drugs. It was also an era of cheap energy. When I married and moved into my first home, I knew about the benefits of insulation, but it was very low in our priorities. Spare cash was put into furniture, decorations, carpets, and so forth. Insulation was invisible and costly, so we disregarded it. It is also doubtful that the cost could be recouped in anything like a reasonable time frame, because energy prices were so low.

How different things are today! For example, there are government offers of financial assistance toward the cost of upgrading properties. If these had been offered at the time, they would have been a powerful incentive for my wife and me. Although they tend to be aimed at low-income people, they are well worth checking out. Details of the Federal Weatherization Assistance Program can be found on www.eere.energy.gov/weatherization. NAIMA also has a website showing incentives for specific geographic areas: www.naima .org/pages/resources/incentives.html.

In considering your own property, it is important to remember that requirements and recommendations embedded in building codes are minimum, rather than *optimum*, standards. And who wants just the minimum of anything good? Installing beyond the minimum is perfectly possible, and you wind up saving more energy and money. More insulation means less energy used. But be careful, if you go too far, it could cost more financially than you are likely to get back.

Let me digress for a moment. Robert Browning, feeling nostalgic for his homeland, once said, "Oh to be in England, now that April's here." Indeed, England can be very beautiful in the spring, but the weather is unreliable. I have awakened in a cold bedroom on many an April morning. On goes my undershirt, and I feel warmer, then my shirt, and I start to feel comfortable. Finally a sweater, and I can face the world—once I've got my trousers on that is! I could put more layers on, but I would only feel slightly warmer and considerably more restricted in movement. The extra insulating effect I get from each extra layer gets smaller and smaller.

Puzzled? So was I when I first started to think about increasing the insulation in my home. The key is to remember that what really matters is how fast heat energy goes out of the building. Think again of water flowing down pipes. If you halve the size of the pipe (i.e., double the resistance), then the rate of flow of water will be halved. Similarly, the rate of heat flow is numerically the *inverse* of the R-value. Technically this rate is known as the *thermal transmittance* (or U-value or, occasionally, the *heat transfer coefficient*). More formally it is defined as follows:

The thermal transmittance or U-value of a structure is the rate of heat flow across unit area when opposite sides are in contact with the air at temperatures which differ by one degree.

When I first moved into my present home thirty years ago, the attic was insulated with 3.5 inches of fiberglass of R-value 11. Over the years our finances improved, and we increased this in 3.5-inch layers to its present thickness of fourteen inches. When I first tackled this job, it also became apparent that there was air infiltration to the attic from around the trap door that was our access point—another case of leaky insulation that had to be fixed at the same time as the upgrade. Table 2.2 shows what happened to the rate of heat flow, as measured by the U-value.

There was a massive slowdown when the first layer was added (above 50%)—just as putting on an undershirt kills the worst of the cold. The improvement bought by the second layer is important but less than that crucial first layer. But look at the slowdown for the last layer—much less. This point is illustrated in figure 2.3.

Table 2.2   Change of Heat Flow with Increased Thickness

Thickness (inches)	R-value (ft² °F hr/BTU)	U-value (BTU/hr ft² °F)
3.5	11.0	1/11 = 0.091
7.0	22.0	1/22 = 0.045
10.5	33.0	1/33 = 0.030
14.0	44.0	1/44 = 0.023

The rate of flow gets less and less as the insulation gets more and more. But as the insulation gets thicker, the extra money saved may not be enough to cover the cost of the extra insulation on anything like a reasonable time frame. The last inch added costs exactly the same as the first and doesn't produce nearly as much savings. What we really need to do is to find the optimum level of insulation for any particular situation.

Saving the Dollars

Determining the optimum level of insulation is, in fact, quite difficult. Assumptions need to be made about interest rates (Would you be better off putting your money in a savings bond?), energy price escalation (What if another hurricane hits the rigs in the Gulf of Mexico?), and the increased resale value of your property. However, the calculation is done routinely for large buildings such as a tower block or factory where it would make sense to hire a heating engineer to "do the sums." If you are good at math then you could learn to do it yourself for your own home—but be warned, it is a lengthy calculation. Personally, I go for a rule of thumb and install 10% above the recommended minimum, reviewing the decision every five years.

Insulation can make a large difference to your energy bills, but it is not the only factor involved. English scientist Sir Isaac Newton, a contemporary of John Locke, discovered that a large temperature difference leads to a high rate of loss. More precisely put, Newton's law of cooling says, "The rate of heat loss is proportional to the excess temperature." The temperature difference between the inside and the outside of a building is therefore another important factor.

We cannot control the outside temperature—unless we move a few hundred

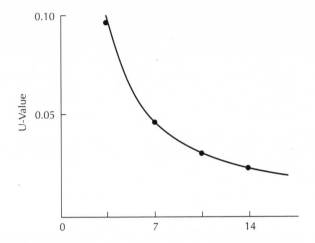

FIGURE 2.3.   Change of U-value (BTU/hr ft² °F) with increased thickness (inches)

miles to the south. Indeed, a neighbor of mine winters in Lanzarote in the Canary Islands, off the North African coast. How I envy her, but, alas, for most of us running two homes is not really an option. But we can control our home's internal temperature by turning down the thermostat by a few degrees (or up if we are in a hot climate). This lowers the temperature difference and saves money. I have seen figures suggesting that a 1°F decrease can reduce heating costs by 3%. Much depends though on the particular heating system and how it is used.

A good plan in a cold climate would be to reduce the thermostat setting a degree at a time until you feel uncomfortable, and then set it a degree or so higher. Don't, however, be like the man in the parable of the man and the donkey. He was worried about how much it cost him to feed the beast in winter. So he reduced its food a little for a week or two, with little apparent difference in the donkey's energy level. It worked so well that he did it again. Then again. Then again. Then one day he woke up to find his donkey dead.

You are unlikely to kill yourself in your quest to save on heating bills, but you can make yourself jolly uncomfortable in the process. Unless you are in desperate financial straights, it just isn't worth it.

The exposed area of a building also plays a major part in its running cost— the greater the exposure the greater the heat energy loss. This is something that architects nowadays take seriously in their designs. Few of them, however, have caught up with the Inuit peoples. The traditional igloo is a hemisphere of snow blocks, and it so happens that a spherical shape achieves the smallest outside area for the greatest internal volume. So heat retention is greatest with this shape. This isn't much help to most of us, though; changing the shape of our homes isn't an option, though we can choose to live in an apartment block where the exposed area is much smaller than in a house.

What's It Made Of?

There's a lovely story in *The Long Winter*, part of the Little House on the Prairie series, in which the family's shack became warmer once they got completely snowed in. How can a layer of thick, cold snow have such a noticeable warming effect? The answer to this mystery is that snow contains large amounts of poorly conducting air. Because the air is trapped, it cannot carry heat energy by convection, making the snow an excellent thermal insulator.

It should be no surprise then to find that commercial insulating materials are designed to efficiently trap air. But there is another way in which heat energy can be lost from a building. When I go walking the hills around my home, I always carry a Thermos filled with hot tea. (Like all Brits, I can cope with any crisis provided I have access to a decent cup of tea.) Have you noticed that the inside of a Thermos is shiny? Radiant heat, which would otherwise escape, is reflected back into it by this radiant barrier. Aluminum foil is used in home insulation for the same purpose. Laying foil on the floor of an attic keeps the radiant heat in, and placing a layer of fibrous material on top traps air and keeps heat from escaping through conduction or convection.

Early Forms of Insulation

Human beings have always needed to protect themselves from the elements, and the history of insulation goes back a long time. A short but fascinating account may be found in Richard T. Bynum's *Insulation Handbook* (2000). I was particularly interested to find that Pliny the Elder in the first century recorded the use of cork as an insulating material for roofs and that mineral fibers from volcanic deposits were used by Hawaiian islanders to blanket their huts.

In the southern UK, houses were traditionally built with thick thatch (a

FIGURE 2.4.   Thatched almshouse at Thaxted, Essex, England

form of straw) roofs. The one shown in figure 2.4 has thatch several feet thick and dates from 1713. It was used by the poor of the parish and today is a private house. Thatch houses are beautiful and are very warm inside during cold weather. Similarly, thick walls of straw bale construction were built in the Sandhills of Nebraska early in the twentieth century. Wood shavings, suitably treated, were also a popular insulation product in the homes of the northeastern United States around the same time.

There are problems, however, with biological insulating materials—they need to be treated with fire retardants and fungicides, for example. And then there is the problem of the vermin that take up home within them! Consequently they are rarely used nowadays.

Today's Manufactured Insulation

Polystyrene. Better known by the brand name Styrofoam, this material was developed by the Dow Chemical Company in the 1940s (www.dow.com/ styrofoam). It fascinated my friends and me when we were children. Its pure whiteness meant we could break it into pieces to make "snow" in the middle of summer. The ease with which we could lift great chunks of it also captured our imaginations. It didn't take us long, for instance, to realize that many of the "rocks" in cowboy films were simply painted polystyrene. Styrofoam has a closed cellular structure within which air is trapped, and it can be either molded or extruded to the shape required. This structure makes it lightweight and water resistant as well as a good insulator.

You can appreciate just how good an insulator it is when you drink from a disposable polystyrene drinking cup. What a boon these are, especially when dispensed from a machine. The drink is kept piping hot and can be lifted without burning the fingers. For building insulation, polystyrene is manufactured into large boards. Trouble is, they are stiff and difficult to cut, making them hard to maneuver in confined spaces such as an attic. Accessibility is not a problem, however, at the construction phase of new buildings, in which these boards are widely used.

Did you know that the beads in beanbag chairs are small polystyrene spheres? They too are used for insulation by simply being emptied into the space to be insulated. Materials used in this way are called "loose fill." Their downside is that they can fall down between gaps or blow around if a draft is present.

Fiberglass. This material first became available in the 1930s. It consists of fibers spun from molten glass and loosely bonded together with resin. These fibers form interlocking cells to trap air. Fiberglass is available as rolls of blanket material or as stiff board. Its R-value varies with density. The tighter it is compressed together the more difficult it is for the heat energy to get through—a bit like trying to push your way through thick undergrowth.

Be careful when handling either form of fiberglass; the fibers often have sharp ends like broken glass, which makes it a skin (and lung) irritant. There is an amusing story in *The New Science of Strong Materials,* by J. E. Gordon (1968), from the early days of this kind of product: "During the days of clothes rationing and before glass cloth was widely familiar I had a large roll of such cloth stolen from the laboratory, no doubt to be made into underclothes. Since glass-fibre is an irritant to the skin I watched the women employees, over a considerable period, to see if they scratched themselves. However, either they were all innocent, or else possessed of great self-restraint, for I never caught anybody" (p. 173).

Mineral Wool Fiber. Similar to glass fiber but sometimes more expensive and less readily available, mineral wool fiber consists of interlocking and lightly bonded mineral fibers and can be supplied in blanket, board, or loose-fill forms. A particularly interesting form is rock wool, which can be spun from the waste byproducts of metal refining. An interesting example of recycling!

Cellulose Insulation. This form of insulation also makes good use of waste material. It is made from wastepaper, such as used newspaper and boxes, shredded into small particles of cellulose fiber. Chemicals are added to provide resistance to fire and vermin. It is only available as loose-fill material.

Vermiculite. A mineral found in natural deposits in many parts of the world (including the US), vermiculite has the remarkable ability to expand to many times its original volume when heated, trapping air within it. It is supplied as a loose-fill material and is very effective and easy to use. However, you may have heard stories of vermiculite being contaminated with asbestos. The US Environmental Protection Agency has a relevant website (www.epa.gov/asbestos/pubs/verm.html#uses) that states: "Vermiculite ores from some sources have been found to contain asbestos minerals, but asbestos is not intrinsic to vermiculite, and only a few ore bodies have been found to contain more than tiny trace amounts." The vermiculite mine at Libby, Montana, was one found to contain contamination. But it has been closed since 1990, and asbestos is not found in measurable quantities in any currently used deposits. More details are available at www.vermiculite.net/, and the EPA website above.

Perlite. Similar to vermiculite, perlite is a granular, lightweight, loose-fill material quarried mainly in the western United States. Like vermiculite, it has to be expanded, and in the process, insulating voids are formed within it. Both vermiculite and perlite are noncombustible, odorless, and resistant to vermin.

Polyurethane. As a young physics teacher, one of the tasks I enjoyed setting for pupils was to go home and list the number of pieces of equipment in their house that contained an electric motor. They were surprised to find that twenty was about average. Had I been a chemistry teacher, an equivalent task would have been to count the items containing polyurethane. As well as its use for automobile tires, shoe heels, adhesives, and so forth, polyurethane can be made into rigid and flexible solid foams. Both forms contain trapped gases, making them excellent insulators. As an added bonus, polyurethane boards can be coated with reflective foil to increase their R-value. Polyurethane is flammable but is usually located behind gypsum wall or ceiling boards that protect it.

Our village Methodist church was built in 1782 and, as a historic building, is a considerable and constant worry for the church authorities. One problem is the roof, where there is a very real fear of tiles being blown off in a storm or simply slipping off as the restraining nails weaken with age. As I write, polyurethane spray foam is being considered as a low cost alternative to re-roofing. The foam is sprayed on from the inside of the roof space, and it adheres to pretty much anything. It is an excellent insulator and also forms an effective air seal, preventing too much air from infiltrating the attic. This solution works equally well in a house.

Polyisocyanurate. This alternative to polyurethane is used in much the same way. It is water resistant (though not recommended for use below ground), and like polyurethane is flammable.

In 1969, when astronauts Armstrong and Aldrin landed on the moon, I was a young teacher in a secondary school in Uganda. I admired the astronauts' adventurous spirit and marveled at the technology involved in their mission, but I was surrounded by abject poverty and wondered if the money could have been put to better use. I needn't have worried. One of many spin-offs of the space program was the development of new insulating materials. A new form of insulation called layered insulation based on space suit technology has recently become available. One such product is called Tri-Iso Super 9 and is manufactured by Actis SA of France. It consists of fourteen layers incorporating wadding, reflective films, and closed-cell foams, yet only has a thickness of about an inch. It is claimed to be equivalent to eight inches of traditional insulation. It is flexible and can be cut to size with ordinary scissors and fixed with staples or glue. Getting close to an ideal insulation product!

There are other less common materials available. Those manufactured by members of NAIMA are all acceptable, and the association's website is a good starting point for finding further details www.naima.org.

Environmental Worries

There is no doubt that if enough homeowners installed proper insulation, we could extend the time before fossil fuels become scarce and reduce the rate at which carbon dioxide is emitted into the atmosphere. But there are other environmental issues involved too. Everything we use, including insulation, comes from somewhere, is processed, transported somewhere else where it may be further processed, is used, and finally disposed of. All of these stages have environmental impacts of one sort or another.

John Wesley many years ago was troubled by the misuse of money and championed three goals: "Gain all you can, save all you can and give all you can" (sermon, "The Use of Money," 1744). Wesley was the founding father of the Methodist church, and to this day Methodists take their use of money very seriously indeed. But be that as it may, Wesley was making the point that you can neither save it nor give it unless you have it in the first place.

Energy is used in the manufacture and transport of insulation; thus, it takes energy to save energy. Like money, therefore, you cannot save it unless you have it in the first place. The total energy used in getting the final product installed in your property, from mining (or growing) the raw material right through to its satisfactory installation, is called its *embodied energy*. Only in the last twenty years or so has this been studied in any detail, but the outcome is encouraging. What has emerged is that the energy conserved far outweighs the energy used in producing the insulation materials described here. You may, however, wish to install the one with the lowest embodied energy in your home. Your choices can be simplified into three groups, which arranged from lowest to highest are:

1. insulation derived from organic sources, such as cellulose
2. insulation derived from naturally occurring materials such as fiberglass, rock wool, vermiculite, and perlite
3. insulation derived from fossil fuels such as polystyrene, polyurethane, and polyisocyanurate

Embodied energy should not be your only concern when selecting insulation, however. Many people really enjoy the warmth of the sun for long periods (not me, though—my skin fries and peels!). If you do, then you should heed government warnings about the increased risk of skin cancer. See, for example, www.epa.gov/sunwise/. Indeed, some research suggests that, based on current trends, 20% of all Americans can expect to suffer a skin cancer at some time. The risk arises from the reduction in the protective ozone layer, which filters out much of a harmful type of ultraviolet light thought to be a cause of skin cancer.

In the past, foam insulation products have used *chloroflourocarbons (CFCs)*, a group of chemicals implicated in the destruction of atmospheric ozone, as blowing agents. Nowadays less potent *hydrochloroflourocarbons (HCFCs)* are used instead, but ideally an entirely non-toxic agent should replace them. One option is a special kind of polyurethane called *polyicynene* (sold under the brand name Icynene), which uses a mixture of relatively benign carbon dioxide and water as the blowing agent. Availability and prices in your area are worth checking out.

A further concern still is that of resource depletion. One of my earliest memories is of my mom and the ritual of "putting out the salvage" every Tuesday. At that time nothing was thrown away that might be reused or recycled, and it was our duty to separate from our garbage anything that might be useful. It was wartime, of course, and Britain was an island cut off from the rest of the world, from where many of our resources traditionally came.

After the war the salvage habit was quickly relinquished, until it started to come back in the early nineties under the impetus of environmental concern. It uses less energy to reprocess recycled material than to process virgin material, and it increases the time it will take for our natural resources to be depleted. Recycling is now becoming almost as common as it was in my early years.

Many types of insulation products contain recycled materials. Cellulose is the winner, for it is made almost entirely from waste paper and card. Likewise, rock wool is made largely from industrial byproducts. Fiberglass also contains around 30% recycled waste glass. Even petrochemical-based insulations (polyurethane, polystyrene, polyisocyanurate, etc.) are in on the act—polystyrene, for example, can use as much as 50% recycled resin. This is a trend we can expect to continue.

IT IS DIFFICULT TO GIVE OVERALL ADVICE about which insulation material is best, for many factors come into play, of which their environmental impact is only one. The trends, though, are clear enough. Insulation standards will continue to rise, and, in my view, environmental concerns will become more dominant. Choice is an individual matter based on what is important to you, but I hope that I have at least identified the major choices that are available, and what their various pros and cons might be.

I hope the following proves to be a useful summary of this chapter. Chapter 3 will address how these materials are used to insulate different parts of a building.

Box 2.1. Summary of Insulation and Insulating Materials

Insulation

- saves money with payback periods ranging from a few months to 10+ years.
- saves energy, even when embodied energy is taken into account.
- reduces volume of external noise.
- reduces condensation.
- increases comfort.
- must be correctly fitted to eliminate unnecessary gaps through which heat energy can escape.

Factors such as climate, building design, budget, and environmental impact are all important issues when selecting insulation for your home.

Insulating materials

- are available in different forms, such as blanket, board, and loose fill.
- have their effectiveness measured by R-value.
- are more effective the thicker they are, but adding more may not be cost effective.
- may lose effectiveness when wet, making moisture control important.
- have differing environmental impacts—organic insulation being least likely to give extreme adverse effects.

CHAPTER 3

Wrapping Up Warm

Insulating Your Home

Over Your Head

C. Northcote Parkinson, in his book *Parkinson's Law* (1958), famously wrote, "Work expands so as to fill the time available." I have a similar theory that rubbish expands to fill available space. This is certainly true of our attic, which was gloriously empty when we moved in and is now packed with a lifetime's clutter. Yet the main purpose of the roof is *not* to form a useful, out-of-the-way storage area. It is to keep the rain out and the warmth in, which means it must be insulated.

The roof, in fact, is usually considered to be the top priority for insulation. If I climb from the basement to the top floor of a house on a cool day, I can feel the difference in temperature. The top floor can be just fine when the ground floor rooms need extra heat. This is because warm air rises due to convection, and the energy is transferred through the ceiling by conduction into the attic, where it escapes through the roof to the outside. This loss can be as great as 20% of the heat energy supplied to the house. Roof insulation is relatively cheap, and the time to recoup the initial outlay (i.e., the payback time) is often three years or less.

FIGURE 3.1.   Fitted blanket roof insulation

Sloping Roofs

The construction of the roof determines the method of insulation. A home with a sloping roof has thick, strong beams of wood called joists running from side to side across the attic space. These are joined to similar beams called rafters to form a series of triangles. The roofing materials (generally tiles or slates) are fixed to the outside of this structure and a waterproof membrane (such as roofing felt) between the tiles and rafters makes the whole construction waterproof. Plasterboard is attached to the underside of the joists to form the ceiling of the room below.

A section of my attic is shown in figure 3.1. The black roofing felt is clearly visible between the rafters, but the joists have been covered by low-density fiberglass insulation to a depth of fourteen inches. Notice, however, that it does not go right up to the edges. The attic is vented from the outside through vents in the *soffit* (the finished underside of the eaves), to form a passageway for air to flow in and out. If these were to be blocked then condensation would be likely, particularly as the space would be much colder after insulating it.

The central part of the attic, not shown in the figure, has been boarded by fixing flooring to the joists to allow the space to be used for storage. There the insulation is only three inches thick—the thickness of the joists. Fortunately it is not drafty, so loose-fill materials could have been an alternative, provided all gaps in or around the ceiling had been sealed first to prevent the materials falling through them. Board material would not have been possible as a retrofit, because the boards are too big to go through the access hatch without extensive cutting. I haven't yet laid foil to reflect radiant heat back down into the house, but I have it in mind as one of my next jobs. This should make a noticeable difference.

The trouble with turning roofing space into a storage area is that it encourages clutter to breed. My son, Andrew, has gone further and converted his attic into an extra bedroom for his growing family. He insulated it by placing the layered insulation Tri-Iso Super 9 between the rafters before covering them with plasterboard sheathing. It has proved an effective solution. Cathedral roofs are often insulated in the same way.

Flat Roofs

But what if you have a flat roof? A false ceiling can be built underneath, and the space between the new and the old ceiling filled with insulating material. Another option is to lay insulating boards or spray polyurethane foam on top of the existing roof and place a permanent waterproof covering on top of the foam. Mind you, I have known people who have had a flat roof and found that whenever it rained, the water would pool and seep through. Long-term waterproofing is not easy to achieve and can be costly. Why any Brit house has a flat roof beats me!

Insulating an attic is usually within the capabilities of the home handyman. If you feel like tackling it, then the above introduction is a good starting point in planning the work, but do look at the various options in more detail. The Oak Ridge National Laboratory has some excellent free booklets and handouts aimed at ordinary homeowners; these can be downloaded from www.ornl.gov/sci/roofs+walls/.

Holding Up the Roof

I once asked a group of students what walls are for. Their answers were revealing. Some saw them as providing privacy. Others felt protection was the top priority. Still more thought of the need for shelter from wind and rain. One or two emphasized an attractive external appearance. Hardly any saw them as necessary to support the roof! This vital purpose can be achieved in two ways. In the UK the most common type of wall construction is masonry—bricks, blocks, or stone are strong enough to carry the weight of everything on and above them; in the US, a timber frame usually carries the weight of the roof.

There is no doubt that timber frame buildings are quicker and cheaper to erect, and some experts believe that masonry has had its day. Nonetheless masonry is more fire- and impact-resistant, leading to lower insurance costs and needing less maintenance—saving both time and money. You don't need to worry about wood-eating termites either, as you would in some parts of the South and Southwest of the United States.

The type of construction affects the way walls are insulated—but what sort of priority should we give to insulating walls? On leaving university, I moved into a small house. The walls allowed more heat energy to flow through them than the roof, mainly on account of their larger exposed area. However, when I did the calculations, I found that although insulating the walls first would save more energy, the higher cost made the payback time much longer. The roof was therefore my priority. This is likely to be the case for most conventional houses.

Timber Frame Buildings

Wood-framed buildings have a long and interesting history. Shown in figure 3.2 is a house in Stratford-on-Avon, England, proudly flying the Stars and Stripes. It is known as the Harvard House and is owned by Harvard University. The university was established in 1636, when John Harvard left his library and half his estate to the new Harvard College in Cambridge, Massachusetts. Born in the UK, he was a contemporary of and knew the "Bard of Avon," William Shakespeare. As a Puritan Christian minister he felt he had to leave his home country to live in the New World, where there was greater religious tolerance

FIGURE 3.2. "The Harvard House" in Stratford-on-Avon, a typical seventeenth-century English "black-and-white" house

FIGURE 3.3.   Exposed panel of a seventeenth-century Tudor house, showing
the wattle and daub

and freedom. The Harvard House was his grandparents' home, and it is interesting to consider its construction.

The house is built on a timber frame of large oak beams pegged and jointed
together. Conventionally these were painted black. Tree laths were fitted in
the rectangles between the beams to form a grid known as the *wattle* (see figure 3.3). Finally a mixture of clay, dung, and straw was daubed over the grid.

The excess was smoothed off and the grid painted. Not surprisingly, this is known as *wattle-and-daub* construction.

Beneath the smooth outer finish, plenty of air is trapped within the daub, making these buildings well insulated and warm. In wattle-and-daub construction, the interior walls usually received a lime plaster coating made from lime and sand reinforced with animal hair or plant fiber. It could be smooth trowelled, left rough, or *parged* (built up into a pattern), but much depended on the status of the owners. People who could afford to hung a cloth painted with a pattern from floor to ceiling—the forerunner of modern-day wallpaper. An additional layer of air was trapped between the wall and the cloth, insulating the house even further.

A modern wood-framed building is a development of wattle and daub. It is typically a 2×6 wood-framed house with insulation, vapor barrier, drywall, and paint on the inside, and sheathing on the outside, as shown in figure 2.2. You may know it as a stick-frame house—though to my mind they are pretty big sticks! The final appearance is determined by the siding, which is fixed to the studs, facing the outside of the house. Two options are to hang tiles on battens fixed to the studs or to build a masonry wall in front of the sheathing, again fixed to the studs with wall ties, but leaving a gap for air circulation and prevention of rain penetration.

There are many different variations available if you are fortunate enough to be able to have a new home built for you, but they must, of course, comply with local building codes. The primary function of standard wood sheathing fixed to the outside of the studs is to give rigidity and strength to the framed structure. However, the studs themselves allow significant amounts of heat to be conducted through them, and increasing timber sizes to allow more insulation is not cost effective. In response to the demand for better wall insulation, foam insulated sheathing is now available for use. Not only does this provide an additional R-value of 2–3.5, but it also protects against condensation by keeping the internal structure warmer. Options are more restricted for later upgrading, though much depends on the particular construction and climate.

The US Department of Energy's website has further information: www
.eere.energy.gov/consumer/your_home/insulation_airsealing/index.cfm/my
topic=1140.

Structural Insulated Panels (SIPs)

We got our first TV in time to watch the coronation of our Queen in 1953, when I was 11 years old. Prior to that, I spent many a happy hour with Dad making models on our kitchen table. When we were bored of this we sometimes practiced card tricks. Dad was particularly good at building structures from a pack of cards. They were very flimsy and fragile of course—but what a sense of achievement when we actually made a structure that both stood up and looked good. And what frustration when Mom opened a door and the draft blew them all down!

You couldn't, of course, build a real house out of cards—though it wouldn't surprise me to find that some people have tried. But what if the cards were replaced by thick factory-built panels? Effectively this is what structural insulated panels are. They consist of two outer skins and an inner core of an insulating material to form a unit. The insulating core and the two skins are insubstantial in themselves, but when pressure-laminated together under the right conditions, they form a composite panel that is much stronger than the sum of its parts. When engineered and assembled properly, a structure built with these panels needs no frame or skeleton to support it—a frameless house!

Structural insulated panels are becoming more popular on account of their high R-values, but they can only be used in new houses. They are certainly well worth considering, but personally I tend to be an inveterate worrier. I would be concerned both about the fire risk and the impact resistance compared with that of my existing masonry house. What if a runaway vehicle slammed into its side? An irrational fear perhaps, but a very real one.

Masonry Walls

Around part of the perimeter of my garden is a dry stonewall—stones skillfully placed so they hold firmly together without the use of mortar. In the field on the other side of the wall, Jim, a local farmer, grazes horses. One day one of them knocked part of my wall down. Jim and I had words and then decided to work together to rebuild it. Dry stonewalls have two outer stone layers separated by a gap (the cavity), which is filled with rubble as the building proceeds. Masonry houses are built in almost the same way, but using mortar and without the rubble filling. In heavy driving rain, water can soak

FIGURE 3.4.   Masonry wall under construction

sideways through brickwork, like tea being absorbed into a partially dunked biscuit, an effect known as *capillary action*. The cavity prevents water from reaching inside the house; any moisture collected within the cavity disperses harmlessly into the atmosphere.

Insulation can be placed within the cavity as the wall is being built, and provided this is done correctly, moisture will not be carried across the gap. Fiberglass or polystyrene board are both suitable materials. Figure 3.4 shows a masonry wall under construction with an outer leaf of brick and an inner leaf of cinder blocks. Modern cinder blocks are made of concrete and cinders but with large voids containing insulating material, which makes them very lightweight. Careful design ensures that they do not weaken the wall, making it structurally unsafe. There is little point in being warm if the roof is liable to fall on your head!

The two leaves are held together by brick ties—the pieces of wire protruding from the blocks in the photograph—and polystyrene board has been placed in the cavity between the two leaves.

For existing cavity walls, loose-fill materials (e.g., polystyrene spheres) or a foamed-in-place polymer foam (e.g., polyurethane) may be blown into the cavity. These require the contractor to pre-drill holes in the wall and neatly fill them on completion. The whole process takes about a day and causes little disruption to the household. It is, however, a professional job, which should only be undertaken by a reputable firm.

Solid Walls

But what if you have solid walls? These have a lower R-value than cavity walls, so additional insulation is highly desirable. Solid timber walls can have an extra false wall constructed to provide a cavity. Almanzo did this in *The Long Winter* by Laura Ingalls Wilder:

> He had set a frame of two-by-fours a foot from the end wall. Now he was sawing boards one-by-one and nailing them to the frame. The rasping of the saw and the hammering were hardly louder than the blizzard's noise.
>
> When he had built the inner wall up halfway, he took out his jack-knife and ripped open a sack of his seed wheat. He lifted up the hundred and twenty five pound sack and carefully let the wheat pour into the space between the new wall and the old one. (p. 170)

His motivation was more to hide his wheat from jealous (and starving) neighbors than to insulate the house, but Almanzo's wall would certainly have kept his home warm!

Something similar can be done with solid masonry walls—a process known as *dry lining*. Thick strips of wood are fixed to the wall and covered over with gypsum plasterboard, and the newly formed cavity is filled with insulating material. The plasterboard is skimmed smooth and the wall decorated to make a well-insulated but slightly smaller room. This works well but is disruptive and time consuming, because wiring and pipes may have to be moved.

Exterior Insulation and Finish Systems (EIFSs)

I've always fancied myself a bit of a home handyman. All the menfolk in my family were skilled workers of one sort or another—except for me. I wound up in university and emerged as a teacher. So at the end of the working day, one way I relax my brain is by doing some sawing, hammering, and banging. But alas, these skills are not inherited, and I've finished up in trouble on more than one occasion. The incident that stands out most is when I had a go at plaster-ing a patch of wall. The wall looked lovely when I went to bed. It wasn't so nice when I got up next day to find a good deal of the plaster on the floor. The secret is building up skills gradually and knowing your limitations.

One job I definitely would not undertake would be installing an exterior insulation and finish system. An EIFS typically consists of insulation board with a water-resistant base coat reinforced by mesh for extra strength. An at-tractive durable finish is added to its outside. The appeal of the system is that it has a high energy efficiency combined with virtually unlimited design flex-ibility—although a stucco finish is the best known.

An EIFS is not designed to be load bearing, and can in theory be fixed to any type of exterior wall, though there wouldn't be much point in adding it to a SIPs construction, which already has similar features. The problem is that it needs a perfect seal, otherwise water can get trapped behind it, leading to problems of rot and mold growth. This is not such a problem with a masonry construction, but you don't need much imagination to see what could happen if incorrectly applied to a timber frame house. It is a job needing a skilled and reputable contractor. Too big a job for the likes of me!

Under Your Feet

Houses are built by starting at the bottom and working up. The reverse is true for installing insulation in an existing house. Since warm air rises, most heat energy escapes through the roof and comparatively little (about 10%) leaves through the floor. Therefore, floors are tackled last.

For a solid slab-on-grade floor, the simplest solution is to lay insulating ma-terial over it, with protective hardboard on top. If the floor is of a suspended timber construction, for example, floorboards nailed to joists, then it may be possible to fix insulation under it. It is, however, disruptive, labor-intensive, and expensive, unless there is a basement or crawlspace underneath. It would only be worthwhile for a new building, or if the boards were being replaced for another reason.

You might wonder if a thick carpet, wooden blocks, or cork tiles provide useful insulation. Sadly, they don't. It is true that these floor coverings feel warm to the touch, but detailed investigation shows that they are too thin to give significant energy savings.

If there is a crawlspace or basement below the house floor, fitting insulation is easier. Fitting it in a confined crawlspace, though, can be extremely difficult—unless you are a contortionist. If the house has a basement with a heating system, then insulation is not really necessary, though codes may require it. A common way of installing insulation in this case is to support fiberglass batts from below by metal supports or wire mesh attached to the basement walls so that they lie between, or cover over, the floor joists (the basement ceiling). But be warned: Jeffrey C. May, in his book *My House Is Killing Me!* (2001), remarks: "In approximately one third of the sick homes with exposed fiberglass basement insulation that I have inspected, and in nearly all crawl spaces, moisture levels had at some point been excessive, and vast populations of fungi and mites were growing in ceiling fiberglass insulation that looked clean" (p. 130). It is clearly a case of alleviating one problem (energy loss) while creating another (health hazard). May wisely recommends a noncombustible drywall covering in these circumstances, but sadly these are rarely installed. The US Department of Energy is, in fact, now recommending sealing and insulating the outer walls of basements and crawlspaces rather than their ceilings. These health problems should therefore disappear over time.

Pipes and Ducts

All modern houses are serviced by the local utilities, including gas, water, electricity, and telephone. Consequently paraphernalia related to the services they provide runs through the building. Wiring obviously does not need thermal insulation, though you should keep a note of where it is for future repair and maintenance. Don't do what was done in my home. The previous owner put a wooden floor in the attic to more easily use it for storage. This covered the wiring, so when I wanted to install extra lighting, I had to lift much of the floor to find where to make the new connections.

By contrast, pipes and ducting must be insulated. Also remember that whenever an unheated attic in a house with a sloping roof is insulated, it ends up colder, since it is the attic "floor" that is insulated. Energy that was passing through it is now restrained by the insulation. So if water pipes pass through the attic, there is a danger of the water freezing. This is bad news indeed, for burst pipes and plumber's bills are never welcome.

The water-carrying pipes in my house have been wrapped with insulating material tied firmly in position. Personally, I find this rather unsightly and nowadays would have chosen to fit split polyethylene tubing around them. Both cold (to prevent freezing) and hot (to retain heat energy) pipes need to be insulated. And, if such pipes go through the roof space, they should, where possible, have the insulation laid over them.

Similarly, ducting should be insulated, or you can lose up to 60% of the heat energy before the warm air reaches the exit register — imagine what that means in wasted dollars! Ducts that pass through unheated spaces will be the most cost effective to insulate, and these should be done first. An inch of fiberglass held in place by vinyl wrap is adequate. When ducts pass through heated spaces, there will only be a small temperature difference between the inside and the outside of the duct, so losses will be small, except when the system is warming up and the rooms are cold. Thus, you will still get useful, but not spectacular, savings from insulating them. Better still, for a new home or extensive remodeling of an existing one, specify a layout that takes the ducting only through heated spaces.

An eye should also be kept open for leaking ducts. Ducts are not like pipes, where any leakage is only too obvious. Cold air leaking in or warm air out (or vice-versa for a duct based cooling system), is going to raise those bills higher than they need be. And unfortunately, it is not unusual to find sections that should be joined that have come apart, or small holes where corrosion has occurred. Sheet metal ducting should be mechanically screwed together at all joints and sealed with approved tapes or mastic.

There is also a slight danger of life-threatening carbon monoxide entering the ducting system, particularly if there are blockages anywhere in the flue. For this reason, and because of the difficulty of balancing the airflow to achieve the right amount entering each room, it is best to seek qualified help for any major changes or repairs to a duct-based system.

Although I have described duct systems from the point of view of warm air passing through them, exactly the same principles apply if they are carrying cool conditioned air. Having gone to a lot of time, trouble, and money to produce cool air in the first place, it's plain daft to lose lots of it on its way to where it is used.

What about the Windows?

I live in a 1970s bungalow. Our main living room has a picture window, approximately seven feet high by twelve feet wide (eighty-four square feet). This provides a beautiful view over the nearby moorland and allows in plenty of light to give us a bright cheerful room. However windows lose more heat energy per square foot in winter and gain more in summer than either walls or roof. It is difficult to quantify the actual amount, but I am told that in the colder parts of Virginia, each square foot of single glazed window loses on average the equivalent of one gallon of oil in a heating season — and it would not be unusual to have three hundred square feet of glazing even in a modestly sized house. At today's energy prices, this is a significant amount. Think of what the figure would be if you lived in Washington State or Alaska.

However, windows, like walls, are multifunctional, and other factors than cost also matter. Installing double-glazing enhances security as well as increases the R-value. Another benefit is soundproofing — remember the sleeplessness of my daughter-in-law Ann, caused by the sheep a-baaing and the owls a-hooting? This benefit will interest anyone who lives in a city or town house with buses and trucks thundering by.

Before we had our picture window double-glazed, we suffered terribly from drafts. Air cooled by contact with the cold glass of the window fell to the ground and moved across the floor — a convection effect. Double-glazing made the inner glass surfaces a degree or so warmer and so reduced the drafts. It also made condensation less likely, and people sitting near the window felt more comfortable because they lost less radiant heat from their bodies. Thus, irrespective of the energy savings that were achieved, I was more than happy with my investment.

Double-Glazed Windows

Only in the last forty years or so has double-glazing become widespread. For this we have to thank British industrialist Sir Arthur Pilkington, who revolutionized glass making by inventing the float process in 1959. This new cheap glass boosted the double-glazing market. It was not until the seventies, however, that modern sealed units—now standard for new houses—emerged. These units are not just a second piece of glass fixed in the window frame. Like modern walls, they are well-thought-out pieces of precision engineering.

All the houses on our avenue were built at the same time, and all the outside timberwork started to rot pretty much simultaneously—my neighbor reckons that the timber must have been floated across the Atlantic from Canada! At the same time, as cheap glass was becoming available, so too was uPVC (*unplasticized polyvinyl chloride*). This neither rots nor needs painting, and so it was a natural choice for us for the frames. It now dominates the market on both sides of the Atlantic, but wood is still used for historic houses where uPVC would look out of place. Also available are aluminum frames, but they are currently unfashionable, being associated with seventies or eighties technology.

The sealed units that constitute double-glazing are two sheets of glass separated by a plastic spacer with durable seals around the edges (see figure 3.5) to trap the air. It might surprise you that the extra piece of glass, being only 1/8 inch or so thick, has little effect on the heat loss. The insulating effect comes almost entirely from the air, which is an excellent insulator provided it is not allowed to move around. If it moves, thermal currents (convection again) will be set up and carry heat energy across the gap. The optimum gap width turns out to be 3/4 inch or so.

If, on the other hand, it's the rumbling of traffic or noisy aircraft flying over your house that worries you, then the optimum gap width is nearer four inches. To keep sound out, you may need *triple* glazing. In any event, the unit must be hermetically sealed to prevent ingress of water vapor and consequent condensation between the glass sheets. The spacers are also filled with a desiccant, such as silica gel, to absorb any residual vapor from the manufacturing process.

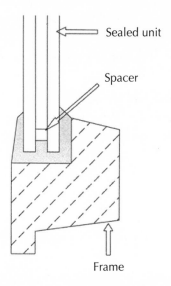

FIGURE 3.5.   Double-glazed window unit

The effect of double-glazing is to increase the R-value from about R-1 to R-2, which sounds like a lot, but is rather like the difference between a nickel and a dime. A big improvement, but no big deal. Further improvement is gained by tackling the heat lost through windows by radiation, but not by introducing a layer of aluminum foil as is done in attic insulation. Technology comes to the rescue again: special glass, known as *low-emissivity glass*, is available. It has an ultra-thin, transparent, metallic coating that acts like aluminum foil by reflecting heat back into the room. This increases the R-value to R-3.

Removing the air from between the units of the window and replacing it with special gases such as argon or krypton increases the R-value further to around R-4. If you have really deep pockets and live in the cold northern states, you can even get "super windows" of R-9, which use all of these tricks in specially engineered triple-glazing.

We on the UK side of the pond are sometimes accused of being fonder of dogs than people. Every year in summer there is a furor, reaching the national press, about someone who has left their dog locked up in a closed automobile while they go shopping. A good security measure perhaps, but the interior of the vehicle rapidly heats up and puts the animal under tremendous stress. In fact, onlookers have smashed the window of such a vehicle rather than witness the suffering.

Glass lets in the heat of the sun, which gets trapped inside, raising the internal temperature considerably. This extra heat is known as the *solar gain*, and it can be a boon or a nuisance depending on where you live. In cooler climates, new houses are often built to maximize this gain by controlling the orientation of the building and optimizing the size of the windows. In hot climates, a traditional answer to overheating caused by the solar gain was to shade the window by building a veranda or even growing tall trees nearby. Nowadays, special low-solar-gain glazing is available. This could well be your optimal answer, but much depends on the specifics of the particular home and the details of the climate. This is definitely a case where expert opinion is important.

Storm Windows

One of the attractions of the Peak District is the historic small town of Buxton with its Edwardian theater, the Opera House (www.buxton-opera .co.uk/) that annually hosts an international Gilbert and Sullivan Festival. I always try to get to *The Mikado* or *HMS Pinafore* if I can—the only type of opera I understand, I'm afraid. The history of double-glazing does not seem to be well documented, but in researching this book I did find that in 1883, W. S. Gilbert moved into a new house. His diaries tell us that "the house contained a marble court, twelve bedrooms (counting the servants) and four bathrooms. Also there was electricity, a telephone and double-glazing." He must have been a technophile to be such an early adopter of these innovations!

What type of double-glazing Gilbert had was not recorded, but it was almost certainly a form of storm window. These are a second set of panes installed on the outside or inside of the primary windows to provide additional insulation and wind protection. In terms of energy, they have R-values of around R-2, but at a lower cost than sealed units. The lower cost comes about partly because of the simpler engineering and partly because of the relative ease of installation—but this must be done right. Air leakage can lead to the windows fogging and frosting up due to condensation between the panes. They may also need to be removed for cleaning, an unwelcome chore.

Not enjoying chores, I would rather set up a program of installing sealed units over a number of years, rather than install cheaper storm windows that I know would not satisfy me in the long run. But this is, of course, a personal choice.

Other Ways of Lowering the Losses

There are other ways of reducing the energy loss through windows. Fitting them with heavy lined drapes (such as your great-aunt may have had) is one such way. However, these only work at night when the curtains are drawn, and it is difficult to effectively trap air between them and the window. A further complication arises if there is a radiator under the window. In this situation, the best solution is to fit a radiator shelf, letting the curtains rest upon it. This long narrow shelf diverts rising warm air from the radiator from passing behind the curtains and making contact with the cold window surface.

Another efficient way to insulate a window is to install shutters. These are traditionally wooden, but ones made from lightweight insulating board protected by a thin casing are available. Clearly, shutters only work when they are closed. They also do not allow light in or occupants to look out, and they need maintenance. Though traditional in some areas, they would not be my choice.

I once visited my younger son, David, in his cold rented flat. I took him to the home improvement store and bought a window insulation kit for a few dollars. This consisted of a thin transparent film, which we fixed over the interior of the windows with double-sided adhesive tape, thereby trapping air. Gentle heat with a hair dryer smoothed it out to almost give it the appearance of glass. The downside is that film taped to the existing window like this is not exactly durable, as anyone with small children could tell you. Yet payback times can be as little as one or two years. This was a good solution in his situation.

Yet, however good your insulation, it is of little use if you habitually leave windows and doors open, or if it is incorrectly fitted to allow air to leak through or around it. There will simply be an uncontrolled heat loss that could prove very expensive. Nonetheless, it is important to have some ventilation in every building—just as you can have too much you can also have too little. We can end up too muggy as well as too drafty.

Too Muggy? Too Drafty?

If the movement of air in a room is insufficient, it becomes uncomfortably muggy and any odors produced remain in the room. If the house still smells of fish two weeks after you have eaten it, your house needs more ventilation. The good news is that your heating bill will be quite low, for little warm air will be escaping. But wouldn't you be willing to pay more in heating bills if you didn't have to smell kippers? There is also a potentially fatal consequence of inadequate ventilation: carbon monoxide poisoning.

It is a great sadness to me to hear about the unnecessary deaths of unwary students from carbon monoxide poisoning, as happens very occasionally in the UK. Students from tropical countries are particularly at risk, for they may never have operated a gas fire before, and are unaware of the dangers. A few, when they feel a draft, block the ventilation to the room, not realizing that an inadequate air supply leads to incomplete combustion. This leads to the formation of highly toxic, potentially lethal, carbon monoxide gas, with occasional tragic consequences. Therefore, I always try to "work in" a warning during my lectures early in the academic year. Excessive air movements are best dealt with in other, safer ways. But what is excessive?

Air movement through a room is measured as the number of air changes per hour (ACH). This is the number of times in an hour that the air in the building escapes to the outside to be replaced by new incoming air. The American Society of Heating, Refrigerating, and Air- Conditioning Engineers (ASHRAE), requires a whole-house rate of 0.35 ACH. Measurements show however, that for some old, poorly insulated houses, rates can reach 2 ACH or even 3 ACH. Imagine the cost of heating such a house when a volume of air equal to that of the whole interior has to be heated from the outside temperature to the desired inside temperature two or three times *every* hour. Naturally the same applies to cooling. There is no point in losing more nice cool (and expensively produced) air than is necessary.

Properly installed breather membranes, mentioned earlier, are the first line of defense against unwanted air infiltration, but they won't stop that awful draft from under the door. A roaring fire makes things worse: the rush of hot air up the chimney gives a rush of cold air under the doors. My mom was very susceptible to drafts and blamed many ailments on them—stiff necks, aching feet, and even colds and flu all came from sitting in a draft. Unfortunately, when she felt a draft, she tended to stoke up the fire. She would have done much better simply tracing and blocking the source of the drafts and allowing the fire to die down a little.

The first stage in reducing these drafts is to inspect all places in the building where they may occur. Likely places are around doors, baseboards, and window frames. Cheap, quick to fit methods of weather stripping are available, but good quality, permanent solutions are more expensive. Simple draft-proofing measures can yield payback times of less than a year and rarely more than three years, for heating the air up or cooling it down less often requires less energy. After that it's a regular free lunch!

Recycling Heat

Even when draft proofing has reduced ventilation rates to 0.35 ACH, this still represents a lot of heat energy escaping to the outside. Advanced heat exchanger systems, sometimes known as *heat recovery ventilators (HRVs)*, enable the warmth in outgoing air to be transferred to the incoming cold air (or vice versa), thereby recycling energy. As a bonus, they remove any excess humidity —and that lingering smell of fish.

Different manufacturers have different ways of achieving the heat exchange, but most send the outgoing warm air through layers of aluminum plates, on the opposite side of which is the incoming cold air. Condensation can be a problem if outgoing humidity is high and incoming air temperatures are low. Either collecting it in a pan or draining it off through additional pipes deals with it. Since the temperature differences are small, the energy flow across the plates can be slow. The flow rates therefore have to be carefully controlled by a fan, and the ducting has to be relatively large, to allow a reasonable amount of heat to be exchanged.

In general, the higher the fuel cost then the more a HRV makes sense, but only if the building isn't losing masses of warm air in other ways.

Table 3.1   Methods of Insulating

Building element	Methods of insulating
Roofs	
Pitched	Blanket, board or loose-fill materials between joists
Flat	Create an insulated false ceiling inside or new layers on top of existing roof outside
Walls	
Masonry	Cavity fill or dry lining of internal walls
Timber frame	Cavity fill, insulated sheathing, batts between studs
Structural insulated panels (SIPs)	Suitable for new houses only
Exterior insulation and finish systems (EIFSs)	Needs skilled installation to existing exterior wall
Windows	
Sealed units	Specially engineered units replace existing panes
Shutters, or heavy drapes	Only effective at nighttime, when closed
Storm windows	Additional panes fitted over existing ones
Floors	
Solid	New layers on top of existing floor
Suspended wood	Insulation fixed beneath
Ventilation losses	
Passive methods	Various forms of draft proofing
Active method	Heat recovery ventilators

Choices

One reason why many people don't look into energy saving in a big way is that there is such a myriad of choices, and therefore decisions to be made, within what is an already hectic life. What would be ideal would be to provide an optimal, universal design standard for everyone. Unfortunately, this is not possible, for what is optimal in Atlanta would not be optimal in Colorado or Anchorage.

Nonetheless, I hope that I have convinced you that improving insulation is feasible and important. Not only will it pay for itself in time, it will also go a little way towards saving the earth. Please be aware, however, that this chapter cannot be any more than a guide to help you know where to start looking. I offer table 3.1 as a relatively simple summary of the different ways of insulating. I also give a few insulation tips in box 3.1.

Box 3.1. Top Tips for Insulation

1. Start with attic insulation, followed by exterior and basement walls, and then windows, floors, and crawlspaces.
2. Don't forget to insulate pipes and ducts.
3. Check ducting for air leaks, and repair them where necessary. For all but the most minor repairs, it is advisable to use a qualified professional.
4. Install storm or double-glazed windows, preferably with low-emissivity glass.
5. Use the solar gain. During the heating season, keep the drapes open on the south-facing windows during the day and closed on the north-facing ones.
6. During the cooling season, keep the drapes closed during the day in unused rooms to reduce the solar gain.
7. In hot climates, think about shading the building or installing low-solar-gain glazing.
8. Do not block vents to prevent drafts. Ventilation plays a large role both in providing sufficient oxygen for combustion and in moisture control.
9. Weatherstrip and caulk to fix drafts.
10. If you have completed all of the suggested methods, and you can live with a long payback time, consider heat recovery ventilators.

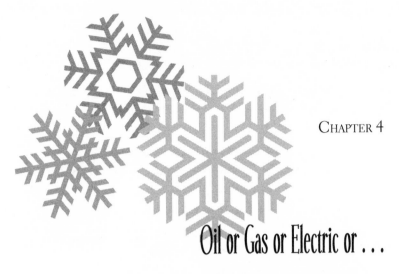

CHAPTER 4

Oil or Gas or Electric or . . .

Fuel Choice and Heating Costs

Pioneer

Brits have a tremendous admiration for the American pioneers. We have this picture of a hard-working people slogging their way across the continent to the Pacific coast, gradually settling the land and trying to wrench a living from it. Imagine yourself on the northern prairies surrounded by a great "sea of grass," as the pioneers were a century or more ago. You find a good place to settle, where you can gather or grow ample food for your family, and store enough to last through the harsh winter months. You build a small house, with its own little fireplace, by logging some trees from the nearby creek banks.

There is plenty of wood available to fuel your fire, but it is not very versatile; it would be difficult to use it for lighting, for example. Fortunately it is zero cost—you don't need to trade your goods with a supplier to obtain a reasonable quantity. Yet it is not very convenient, since you have to cut, chop, and store it, and safety is a major concern: flying sparks are an ever-present hazard to a flammable wooden structure.

As the years pass, the farmstead flourishes and technology improves. Coal and kerosene are both sold in the local township, and so you include a stove in your house to supplement the fireplace. However, these cannot be delivered if the train can't get through, and in winter the creek banks are not safely accessible for supplies of timber. The Ingalls family found themselves in this situation, with tracks snowed under and supplies not getting through, in *The Long Winter*. They twisted and knotted hay into tight sticks and burned them until the late spring eventually arrived. This was an ingenious solution but exceptionally hard work, as the following extract shows.

> Often Pa came to the stove to warm his hands. "My fingers get so numb," he said, "I can't make a good twist."
>
> "Let me help you, Pa," Laura begged.
>
> He did not want to let her. Then he admitted, "But somebody's got to help. It is going to be more than one person can do, to keep this stove going and haul hay for it." Finally he decided, "Come along. I'll show you how." (p. 195)

Later still your home is connected to a distant power station, allowing the choice of electrical heating. But there are other problems. Many scientists believe that our reliance upon electrical power adds up to a huge environmental impact that could change the climate disastrously in the long term. You therefore start to look into other options—your own small wind generator perhaps, or a small hydro generator immersed in the river to which the creek flows.

This is roughly where some people are today. There is wider choice, both of type of fuel and of supplier. Yet the criteria of choice are still the ones used in earlier days:

- availability
- versatility
- cost
- choice of supplier
- safety
- convenience
- environmental impact.

In this chapter we turn our attention to modern day fuels to see how they measure up on these criteria. Remember though that a change of fuel may entail an expensive change of heating system. Consequently, improved insulation may prove a better option, and should be investigated prior to (or alongside of) the choice of fuel.

What's Available?

In the early seventies I went to a college reunion. I thought I would wear contemporary dress from the early sixties, when I was educated, but my old clothes just didn't fit. So I went around gentlemen's outfitters looking for new drainpipe trousers, velvet jacket, and "brothel creeper" shoes. Luckily my bootlace tie still fit perfectly! Alas, the clothes I wanted were not available anymore, except from specialty suppliers at huge cost, so I ended up in a conventional dark suit.

My choice was limited by availability. Most of us do not have the skills to tailor our own suits. Nor can we create a fuel to meet our individual requirements; we have to use whatever is available. Figure 4.1 shows the percentage of US households using particular fuels as their main heating fuel—but do remember there are variations between states according to climate and other local factors. You will need to check availability in your area if you are contemplating a change of fuel.

As shown, the majority of people rely on *natural gas* brought to their home through the extensive pipework of the distribution system. There are areas, though, where the necessary infrastructure does not exist—approximately 27% of US neighborhoods. It may be possible to get a supply, but the cost is high and has to be borne by the consumer. This situation is akin to my dark suit. The "Teddy boy" outfit was both available and preferable, but the extra costs were exorbitant, and I ended up with something else. The criteria for fuel choice, in this case availability and cost, cannot be considered in isolation from each other.

Connecting to the electricity supply is generally not a problem in the US. However, you can go it alone if you wish. Power outages due to the mismatch of supply to demand (California), or from the use of aging equipment (northeastern states) have led some people to explore alternatives. Diesel generators are widespread, though they are more often used as backups for critical sites such as hospitals. Like other small-scale generators, they are expensive and put you in a position like tailoring your own clothes, good only if you have the time, skills, and finances to do it yourself.

Small-scale environmentally friendly generators—wind generators, solar

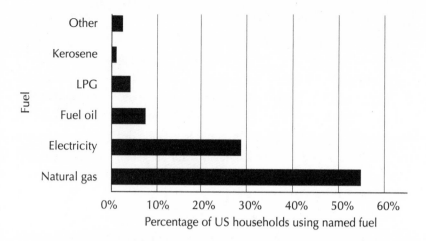

FIGURE 4.1.   Main heating fuel use by percentage of US households.
(Approximately 1% of US households have no heating.) Adapted from
www.eia.doe.gov/emeu/recs/recs2001/ce_pdf/spaceheat/ce2-1c_climate2001.pdf

panels, and hydropower units — are also options, albeit pricey ones. Continuity of supply is a problem: the sun doesn't always shine, nor does the wind always blow, and river flows can be erratic. Clearly the feasibility of these alternatives must be carefully explored before parting with serious money. You will find more detail in Chapter 8. It is also worth taking a look at the website of the National Center for Appropriate Technology (www.ncat.org), if you are seriously considering these options. You may not find all the answers there, but there is an extensive set of links, which should lead you to the information you want. The Centre for Alternative Technology (www.cat.org.uk) fulfills a similar function for the UK.

Adequate storage facilities (and good access for delivery vehicles) are necessary if you use fuel oil or liquefied petroleum gas. Underground storage tanks present potential problems of corrosion-induced leakage into both the soil and groundwater, particularly if they were installed some years ago. Anything over fifteen years old should be considered at risk. You won't be popular if your oil contaminates your neighbors' water supply, and what's more, the cost of the cleanup could easily bankrupt you. There is also the fire risk to consider.

At a minimum, installations must comply with both the local and federal regulations. An alternative to your own tank, adopted in some areas, is to receive your supply from a central neighborhood tank with each home metered separately. The delivery company then takes responsibility for the storage tanks. This service costs the householder extra, but what price peace of mind?

When I was a boy, I lived in a coal mining area and the coal man, unloading from the back of his flat-topped wagon, delivered coal to our home. This was wartime and so the wagon was horse-drawn; trucks used scarce, rationed diesel fuel. Mind you, there are advantages to horse-drawn transport. About once a week, Dad went out "oss muckin" with bucket and shovel. The droppings were collected from the road and ended up on our vegetable patch—it's great fertilizer!

There was never a shortage of coal, however. It came in large sacks and was tipped into the coal shed at the back of the house, from where Dad collected it daily to make the fire. Coal has now declined as a household fuel. The decline is, however, due to the economics of mining and delivering the coal rather than its availability. It remains in the ground ready to be extracted should prices move favorably towards it.

Wood is one of the earliest materials used for heating, and still readily available in some states. *Coppice cutting,* or coppicing, is a method of woodland management by which young trees are cut down to a low level. Many new shoots grow up, and after a few years these are harvested, and the process begins again. The New Jersey Department of Environmental Protection points out that "early settlers in North America used coppicing in the hardwood forests for producing fuel wood and charcoal. Later this method fell from favor as coal and oil displaced wood as fuel. With today's return to the forests as a source of fuel, coppice cutting is a method worth thinking about" (www .nj.gov/dep/parksandforests/forest/community/enviro_silvi.html).

Pelleted wood fuel is also available, sometimes originating from the waste in lumberyards or furniture factories. In the rural area where I live, local farmers from time to time sell logs in bags, which they deliver on request. Don't forget, though, that dry storage is needed, and inquire about future availability before investing in a wood stove.

Table 4.1   Years of Supply of Common Metals

Metal	Years of supply
Iron/Steel	77
Aluminum	162
Copper	22
Lead	10
Zinc	16
Nickel	43

Source: Adapted from Norgate and Rankin 2002

But Will It Be There Tomorrow?

When I went off to my reunion, I bought a new conventional dark suit especially for the event. The decision was mainly driven by availability; however, I also admit that in the ten years since leaving college, I had changed from being trendy to being someone who was happy to be like everyone else. But, there was another issue. I had a wife, young children, and a mortgage, so it was also important that my purchase should last well into the future. Good quality suits worn only for special occasions have lifetimes of fifteen to twenty years, which is about the same as a household heating system. Will my choice of fuel "last" as long?

There is a big problem here: future availability of all minerals, including fuels, is difficult to assess with any certainty and is shrouded in controversy. Table 4.1 shows global estimates of the number of years of supply left for six common metals. These estimates change as more deposits are discovered, as recycling rates improve, and as rates of use rise due to population increase or the continuing industrialization of Third World countries. There is certainly a problem, particularly as the modern equipment we need for keeping ourselves comfortable, is largely made from metals or alloys of one sort or another.

But is the problem as serious as all that? In 1981, Julian L. Simon, an economist and author of *The Ultimate Resource,* saw human ingenuity as the ultimate resource. In this view, supply problems reduce to a contest between resource scarcity and human ingenuity. Simon offered a bet, which was accepted by the environmental scientist Paul Ehrlich, concerning the prices of five metals over a ten-year period. If the prices fell, then human ingenuity was overcoming problems of increased scarcity. There could be a number of possible reasons for this. Human ingenuity could lead to acceptable substitutes being developed, or new methods of extracting the metals being found, or viable technology for recycling them being discovered. Over the decade 1980–90 prices did fall, and Simon won hands down.

As a good Methodist, I distrust gambling and wonder what a bet like that proves. Is a time scale of ten years anywhere near long enough to prove Simon's point that human ingenuity as the ultimate resource is capable of overcoming resource scarcities? Or is there anything special about that particular ten years? An economic recession, for example, could also lead to falling prices. The controversy still rages on. What is absolutely certain is that there are finite amounts of minerals in the earth's crust, and these will run out at some time in the future. Recycling will help, but some of the metal is dispersed and rendered unrecoverable in use. Recycling can delay but not solve the problem.

The same arguments apply to fossil fuels. Gas reserves, at current usage levels, may last for about sixty years, but new discoveries could push this up to around one hundred years. The situation seems worse with oil, with estimates starting at forty years but rarely going above eighty years. There is more coal, with lifetime estimates ranging from two hundred to three hundred years. Depressing as this is, the conventional wisdom now in much of the scientific community, is that the carbon dioxide produced by burning these fuels will produce such drastic climatic change that they will cease to be used long before they run out. The long-term prospects for fossil fuels are therefore not good, and rising prices due to increased scarcity seem inevitable in the short- to medium-term.

But wait; there is some good news. The good news is that electricity will always be available, since it can be generated not only from fossil fuels but also from nuclear materials and renewable sources. Perhaps human ingenuity can rescue our energy supplies after all.

There are problems, however, with the nuclear option. First, current nuclear stations rely upon economically extractable reserves of uranium being available, and estimates start at eighty years or so of supplies left in the earth's crust. And secondly, although many nuclear power stations are operating safely and efficiently at present, the long-term prospects are much more doubtful as some deferred safety issues (waste disposal, plant decommissioning, etc.) become critical.

We Brits may be more sensitive than Americans about the safety issue. Americans were lucky in that the accident in 1979 at Three Mile Island, near Middletown, Pennsylvania, was quickly contained. Not so the Chernobyl disaster. I well remember dangling a Geiger counter outside our laboratory window in 1986 trying to detect the expected radiation as the wind blew some of the fallout over the UK. (I couldn't.) And there is still a Chernobyl children's project based in the High Peak where I live. Children with multiple illnesses often including cancer, probably caused by the continuing radioactivity, are brought over from the Chernobyl area to stay with families and have a holiday. Local people give gifts for them to take back, and medics (local doctors, dentists, opticians, etc.) give them their services free.

Current nuclear power stations rely on nuclear fission. Research is ongoing on another source of energy called nuclear fusion. This would use a process very like that which occurs on the sun. If it can be reproduced on Earth, fusion could solve the world's energy problems. Current plans are to have a demonstration reactor operating by 2035, with the first commercial power stations by 2050. However, the technological challenges are formidable and the research very expensive. My guess is that fusion won't make a significant impact before 2080, if at all. By that time, global warming will be well and truly underway, unless effective measures are taken now to limit its effects.

What then of renewables such as hydroelectricity, wind, tidal, biomass, and solar sources? Currently, these sources (except for hydro and wind power) are at a relatively early stage of development for large-scale use and are not, or are only marginally, economic in most cases. Clearly, the above arguments indicate that they will have to be used extensively in the future. Small-scale generators, suitable for individual homes, are commercially available. We will leave details of them until Chapter 8, in which we indulge in a little crystal-ball gazing.

So, which fuel should I buy? Putting aside the other criteria and thinking just on availability grounds, coal or electricity seem the best options. Though using electrical heating derived from fossil fuels has major problems, because even the most modern thermal power stations have an efficiency of only 40% or so. This is limited by the underlying physics (the second law of thermodynamics) and cannot be improved much more. The other 60% disappears into the atmosphere as heat. On the other hand, if fossil fuels are burned in a furnace or boiler at home, the efficiency can be near 90%. Burning these fuels in power stations rather than in homes hastens their depletion.

But there is more to choosing than availability. Read on.

Versatility

I went off to my college reunion in my new dark suit, admittedly with a tinge of regret, though I knew in my heart of hearts that it was the right decision. After all, a conventional suit could be worn to business events, weddings, or funerals. I could hardly turn up in a Teddy boy suit to such events. Versatility won the day.

With fuels, electricity wins hands down on versatility. Not only can you use it for heating, you can use it for cooling too. There are, of course, no wood-fired air-conditioning units! It is theoretically possible to use natural gas or oil for cooling, based on the technology of old-style gas refrigerators, but the expense and low efficiency of these methods would rule them out these days. Electricity can also be used for cooking, entertainment, lighting, and pretty much everything that a householder could possibly need. Nothing else comes anywhere near.

I would add one small note of caution. I live in the High Peak of Derbyshire in the UK. A couple of times a year we lose power for a few hours, usually because of storms. Going for electricity only on the basis of versatility is fine, but remember: if your power goes then everything else may well go with it. So we cook on gas, have a backup paraffin heater stored in our shed, and have gas-fueled camping lamps easily accessible—but then, I am the kind of guy who wears suspenders in case his belt snaps!

A Gas Guzzler?

The time comes in every dad's life when his offspring wants his own wheels, usually at an age or state of maturity when Dad would fear to be on the same bit of road. I tried the generous approach by paying for the fixed costs of my son

Andrew's car for the first year, letting him know just how much these cost. This left him with only the price of gasoline and maintenance to cover. I hoped he would get something small and economical. He wanted to impress the girls and ended up with a very old sporty car, which needed major repairs within six months. Guess who had to pay for them. (Me, of course!)

Running a heating system is a bit like running an automobile. There are annual (or monthly or quarterly) charges, as well as the cost of the fuel. How much energy you get from the fuel is also important. If you compare the gas model of your car with the diesel model, you will find that the diesel one gives at least 20% more miles per gallon. Finally, what of the equipment the fuel is being burned in? Just as some cars give 15 mpg, others, using the same fuel, may return 40 or even 50 mpg.

The effective cost of a fuel therefore depends on four factors:

- regular charges, such as maintenance and insurance,
- the fuel cost,
- the heat energy available within the fuel, and
- the efficiency of the equipment being used.

Regular Charges

The easiest of these costs to deal with are the regular charges. These include maintenance costs and sometimes a charge made by the energy company to cover the cost of installing and maintaining the supply system (meters, pipes, wiring, etc.). These charges are independent of the amount of fuel supplied and are generally a small part of the overall annual bill for all but the smallest of customers. Few people change fuel because of these costs.

Unit Cost

The unit cost is the price paid for each unit of fuel supplied. These prices can be obtained either directly from the energy companies or from a comparison service such as www.energyshop.com/es/. Comparison between fuels is not straightforward, however, because different fuels are sold in different units—natural gas by the therm, electricity by the kWh, and oil by the gallon. Buying by the gallon is particularly confusing for a Brit in the States or an American

Table 4.2   Calorific Values of Common Heating Fuels

Fuel type	Number of BTU/unit	Number of kWh/unit
Fuel oil	130,000/gallon	41.0/gallon
Electricity	3,412/kwh	1.0/kwh
Natural gas	1,025,000/thousand ft^3	300.3/thousand ft^3
Propane	91,330/gallon	26.8/gallon
Wood	8,000/pound	2.3/pound
Kerosene	135,000/gallon	39.6/gallon
Coal	28,000,000/ton	8,204/ton

Source: Adapted from www.eere.energy.gov/consumerinfo/energy_savers/energyuse.html

in the UK, for the British gallon is bigger than the American one by a ratio of approximately six to five. Yet whatever unit they are sold in, all fuels do effectively the same thing: supply us with energy to keep us warm.

To get some sense of the situation you need the cost per unit of *energy* supplied, because it is the energy in the fuel that is needed, rather than the fuel itself. Think back for a moment to the twisted hay sticks that Laura Ingalls and her father were reduced to burning in *The Long Winter*. This kept them going, but it was far from satisfactory as a fuel: "In the lean-to Laura and Pa twisted the hay until their cold hands could not hold the hay to twist it and must be warmed at the stove. The hay-fire could not keep the cold out of the kitchen but close to the stove the air was warm" (p. 234). As well as burning too quickly, the hay sticks simply did not have enough energy in them. Earlier in the winter, when they were able to burn coal or lumber, the problem did not arise, for both of these contain more energy, weight for weight, than hay. Both hay and wood are mostly cellulose. Had Laura and her father been able to compact the straw to the density of wood, all would have been well. In fact nowadays, you can buy hand-cranked log-makers that will compact your newspapers, which are also cellulose, into handy burnable briquettes—useful if you are miles from recycling facilities (see www .naturalcollection.com/natural-products/Make-Logs-not-Waste.asp). Since these work with hay too, they would have solved Laura and Pa's problem quickly and easily.

Box 4.1. Fuel Prices per KWh

Example 1

A gallon of residential heating oil costs $1.80, and its calorific value is 41.02 kWh/gallon. What is the cost per kWh?

1 gallon costs $1.80

41.02 kWh costs $1.80

1 kWh costs $1.80/41.02 = $0.047

Residential heating oil costs 4.7 cents/kWh

Example 2

Electricity is priced at 9 cents per unit. What is its cost per kWh?

1 unit of electricity is 1 kWh

Electricity costs 9 cents/kWh

Scientists would say that the hay sticks had a low calorific value compared with the wood or coal. The calorific value is the energy available in one unit of the fuel. Table 4.2 shows the calorific value of some common heating fuels, expressed in the units in which they are most commonly sold. Conversion factors for these units have been included in the appendix.

This data helps a little, but what we really want to know is the cost of each kwh of heat energy that comes from each fuel: the cost per kWh of the fuel. A further difficulty is that prices change, and today's price could be slightly different tomorrow. Nonetheless the calculation is fairly straightforward, and if you are interested in the math, box 4.1 shows how to do the sums. A quicker, but less accurate way is to use the conversion factors for the more common fuels given in table 4.3.

For further information about energy conversions and prices, see: www .baywinds.com/new/FossilFuel.html.

The prices given in box 4.1 and table 4.3 have been calculated from price data on various US websites in early 2006. But don't do away with your electri-

Table 4.3   Factors to Convert Fuel Prices to Cents/KWh

Fuel	Quoted unit price	Factor	Comparison unit price
Electricity	9 cents/kWh	1	9 cents/kWh
Natural gas	80 cents/therm	0.034	2.72 cents/kWh
Residential heating oil	180 cents/gallon	0.026	4.7 cents/kWh

cal heating system on the basis of this information. There is yet another aspect to consider before we have the relative prices sorted.

Equipment Efficiency

The final factor in selecting a fuel is the efficiency of the equipment in which it is being used. It is like the difference between my son's old sports car at 25 mpg, and my modest little Ford hatchback at 40 mpg—the efficiency of the equipment matters. For most fuels, energy is lost in extracting it from the fuel and delivering it into the space to be heated. For example, if your gas fire is more than twenty years old, its efficiency could be as low as 50%, meaning that only half of the energy in the natural gas is actually delivered into the room. Most of the rest goes up the chimney, but you still pay for it.

In order to make cost comparisons between differing fuels, we really need to know the cost of each kWh actually delivered into the space to be heated—the cost per useful kilowatt-hour. This depends on the efficiency of the equipment being used. Some different heating system efficiencies and consequent fuel costs/useful kWh are given in table 4.4.

The list of equipment and systems in table 4.4 shows the huge variety available and indicates large differences in performance. Each will be considered in more detail in later chapters, but for now the focus is upon the fuels. If you are interested in the math, box 4.2 shows how the costs for useful kWh are arrived at, based on the data in table 4.3.

Annual Fuel Utilization Efficiency (AFUE) Ratings

The whole thing looks awfully complicated, doesn't it? What we would really like are actual test figures for the specific piece of equipment we are consider-

Table 4.4 Typical Heating System Efficiencies and Cost/Useful KWh

System	Efficiency (%)	Cost (cents/ useful kWh)
Wet (hydronic) systems		
Gas and oil boilers		
Older boilers (more than 20 years old)	55–65	4.5
Modern boilers	60–70	4.2
Condensing boilers	80–85	3.3
Warm air systems		
Gas furnaces		
Conventional furnaces (more than 20 years old)	50–60	4.9
Newer standard furnace	70–80	3.6
High-efficiency condensing furnace	80–90	3.2
Oil furnaces		
Conventional furnaces (more than 20 years old)	50–60	8.5
Newer standard furnace	70–80	6.3
High-efficiency condensing furnace	80–90	5.5
Electric furnace	100	9.0
Individual room heaters		
Gas fires		
Older gas fires (more than 20 years old)	45–55	6.7
Modern gas fires	55–65	4.5
Wall heater with balanced flue	65–75	3.9
Electrical systems		
Baseboard/resistance systems	100	9.0
Storage radiators	100	9.0
Air source heat pump	140	6.4
Ground source heat pump	250	3.6

Box 4.2. Fuel Prices per Useful KWh

Example 1

The cost per kilowatt-hour of central heating oil is 4.7 cents. This is burned in a boiler of 65% efficiency. What is the cost per useful kWh?

1 kWh costs 4.7 cents but only 0.65 kWh of this is useful, therefore
0.65 kWh of useful energy costs 4.7 cents, therefore
1 kWh of useful heat costs 4.7/0.65 = 7.2 cents.
The cost per useful kWh is 7.2 cents.

Example 2

Electricity costs 9 cents/kWh. If it is "burned" in an electric fire, what is its cost per useful kWh?

Electrical fires are 100% efficient, therefore the cost is still 9 cents/useful kWh.

---

ing, rather than the broad figures of table 4.4. The good news is that specific figures are available, for the Federal Trade Commission requires that new furnaces, boilers, and selected other equipment display their annual fuel utilization efficiency (AFUE). This compares the amount of heat energy delivered to the rooms of the house to the amount of fuel supplied to the equipment generating the heat. Technically it is

the ratio of the heat energy output to the total energy input expressed as a percentage.

So an older furnace or boiler with an efficiency of 60% delivers 60% of the energy in its fuel to the room, and the rest is lost, mainly through the flue or chimney. Consequently, the higher the efficiency, the less fuel a piece of equipment will use. For a fuller discussion, see www.eere.energy.gov/consumer/your_home/space_heating_cooling/index.cfm/mytopic=12530. Also, the US Department of Energy has determined minimum fuel efficiencies for heating equipment sold in the US. But note that this is a minimum. You can be sure that the room heater you buy meets this standard, but the actual AFUE rating is also worth checking to see if it exceeds the minimum, and if so, by how much.

Where From?

My friend, Bill, a retired math teacher, decided that he needed a new heating system. As you would expect, this decision was reached after much thought and not a few hours of calculations. He'd been quite happy with his natural gas system, so he called the company for some further information and advice. He was met by an automated system and worked his way through the levels to get to the department he thought he needed. After being kept on hold for ten frustrating minutes, he spoke with an employee who offered to send him some leaflets. He'd also been satisfied with his electricity supplier and wondered what kind of systems they did, though he knew they would be more expensive when it came to running costs — so he called them as well. He was answered after one ring, carefully listened to, and immediately put through to the right department. Most of his technical queries were answered there and then. Offers were also made to send him some leaflets, and for someone to come around to his house to give him individual advice about equipment and costs.

Who got his order? The electricity company of course. Perhaps it would cost him a little extra per year, but quality of service mattered at least as much as money did. The quality of the product was not an issue. He knew that electricity, oil, or natural gas is pretty much the same no matter what company it is bought from, and solid fuels are priced by grade.

For Bill, the main concern was quality of service. For many others, especially those on a low budget, price is the overriding factor. If people in this situation follow the advice in the previous section (in conjunction with a comparison service such as www.energyshop.com/prices), they will not go far wrong. This comparison service also provides historical data to allow comparisons between fixed price deals and standard variable rates. Such deals make budgeting easier, but it is worth remembering that the volatility of fuel prices makes such deals a bit of a gamble — they might not yield savings over the term of the deal.

Price and quality of service are undoubtedly the major factors for most people in choosing their supplier. There is, however, a growing number of green consumers (my son, Andrew, now grown considerably more sensible, is one of them) and some companies offer environmentally friendly contracts. A slightly higher tariff is applied and the extra revenue funds the development of renewable energy. Smaller companies that supply energy only from renewable sources are also entering the market—but their supply may not be available in all areas.

Is It Safe?

My maternal grandpa, Grandpa Walker, was a coal miner in the days when it was a hard, dangerous, and poorly paid job (even more so than today). His lifestyle is described exactly in the novels of his contemporary D. H. Lawrence, who based *Sons and Lovers* in our part of the UK, the Nottingham/Derbyshire coalfield. As a miner, Grandpa had one important perk. He was entitled to free coal for the whole of his life, including after his retirement. But as he got older, the dirt and inconvenience of coal started to bother him. It had to be brought inside the house, ashes had to be removed, and it was sometimes difficult to get the fire going—all daily tasks even in freezing weather. So Grandma and Grandpa changed to gas.

Soon after, however, there was a gas explosion in our town, which flattened two houses. Mercifully no one was killed, but it was enough to frighten the old folks and they changed again, this time to an electric fire at huge expense and inconvenience. They figured that availability, price, and a good supplier are important, but not much use if by keeping yourself warm, you expose yourself to life-threatening dangers.

It's not just the elderly who need to know about the safety aspects of their fuel choice; we all do. Luckily it is quite straightforward, being largely a matter of ensuring that equipment is well maintained and that the correct procedures are followed. For fuels that are burned (coal, wood, oil, and gas), the issues are ventilation, adequate fluing, and prevention of leakage. For electricity, the main issue is protecting against faulty equipment.

Ventilation

There are certain times in life that stick in your mind forever. For me, one of these was my first real chemistry lesson, but for all the wrong reasons. We spent over an hour on how to light and use a Bunsen burner. Boring or what? It's a wonder I ever became a scientist! But for many things we learn at school, it is often some years before we realize their full significance.

With the Bunsen burner, the significant bit was what happened when we turned the collar at the bottom of the burner. This exposed a hole where the flame roared excitingly. The flame was almost invisible except for a blue cone sitting on top of the tubular burner. This, I learned, was the hot flame when all the gas was burning. When the hole was closed, the flame became quiet and yellow, rather like a candle flame, and blew about with whatever slight draft there was in the laboratory. This, I learned, was the cool flame. If we touched this flame to a cold saucer, it left a black deposit as a candle does. For safety we always had to turn the burner down to this cool flame when we were not using it, so the flame would be visible.

Years later I learned that natural gas contains carbon, and that all of the gas has to be fully burned to get all of the heat out of it. In a Bunsen burner, this is achieved by supplying plenty of air to the flame, which then burns with a bluish roaring flame. That is what the hole at the base of the burner is for. Under these conditions, carbon dioxide, which is relatively harmless, is released.

However, if the oxygen supply is inadequate (i.e., the hole is closed), then either the flame will not ignite at all, or more seriously, it will burn with a yellow flame due to hot particles of solid carbon (the material that left a deposit on the saucer) within it. This is not in itself hazardous, but carbon monoxide gas, which is dangerous, may also be produced. This gas is colorless, odorless, and highly poisonous. Victims tend to go drowsily into a deep sleep from which they don't wake up. A Bunsen burner releases insufficient carbon monoxide into the air to be dangerous. However, for household gas equipment, a yellow flame is a warning sign of the possible presence of the gas and should be taken very seriously. Carbon monoxide detectors can be purchased to give a warning, but the main line of defense is regular inspection of gas equipment to ensure there is an adequate oxygen supply to the burners. Annual maintenance contracts are available for this purpose.

Exactly the same problem can occur with any carbon-based fuel: coal, oil, wood, or even the hay sticks of *The Long Winter*. It is absolutely essential that burners or grates have an adequate air supply and that the ventilation apertures of gas burners and all airbricks and vents have a regular inspection. So if you see black deposits or a yellow flame anywhere near your heating equipment, then you are at risk. The only exception are the types of gas fire, sometimes called "living flame" fires, which are designed to give a pleasant real flame effect.

One final word of warning. Figure 4.2 shows a grid in the floor of my living room. This supplies combustion air to the boiler, which is situated behind the gas fire in the recess at the bottom of the chimney (known in the UK as a back-boiler). Unfortunately we are an untidy lot in our house, and newspapers or grandchildren's toys often cover the grid. Consequently, I have had an extra grid placed in my wall. I know the dangers of having too little oxygen reaching the burners. Regrettably not everyone does, sometimes with tragic consequences.

These dangers are less likely in the US, where basements are more common. For example, if a boiler or furnace is situated in a basement, there has to be sufficient floor area (as specified in the appropriate building codes) to provide the necessary amount of combustion air. Indeed, for new homes, many codes require an air duct that brings fresh air direct from the outside to the furnace or boiler. The general principle that vents should not be impeded still applies, however. A household air distribution system is a carefully designed piece of engineering, and reducing the flow *anywhere* within it could lead to problems.

Flues

In 1863 Charles Kingsley published his famous book, *The Water Babies*. It is the story of what befell Tom, a young chimney sweep employed by the brutal Mr. Grimes. Nowadays it is often seen as moralistic and sentimental, but the serious issues of enforced child labor and how the poor should be treated are tackled head on. The job of the child sweep was to clear the flue by brushing the internal brickwork of the chimney as they climbed up on the inside. (The flue is the passageway for exhausting waste gases—in this case the hole formed by the brickwork of the chimney.) This was very frightening, and the children were often reluctant to go up. Some masters pushed them into the flue and then lit the fire to force them to climb upward. This cruelty, which no civilized people would tolerate nowadays, is

FIGURE 4.2. Floor-mounted ventilation grid

quite a contrast to the children happily climbing chimneys in the popular film *Mary Poppins*.

But why must chimneys be swept? Apart from the mess of falling soot, originating from the fires, there are also safety issues involved. The soot is carbon, deposited on the inside of the chimney from incomplete combustion of the fuel, just like the carbon deposited on the saucer by a Bunsen burner. The situation is even worse with wood fires, because creosote, a tar-like substance, can be deposited on the inside of the chimney. If this ignites, a nasty fire can result. This is no joke. According to the US Consumer Product Safety Commission, in 2002 there were 26,300 fireplace and chimney fires resulting in twenty deaths and $148 million worth of damage. Flues should be checked and cleaned annually. The situation is perhaps less critical with oil-based heating, but an annual check is still advisable.

A further hazard arises from impurities in the fuels, which emit acidic gases when they are burned. These can condense on the inside of the flue and cor-rode the walls, making them unstable. So bricks as well as soot can fall down a chimney, to say nothing of the child sweeps of Victorian times falling into the fire because of insecure footings. Flues should therefore be insulated all around to stop the gases from cooling too much and condensing. Finally, a stainless steel flue liner can be added, so any condensation that does occur doesn't harm the brick.

Fully or partially blocked flues can also interfere with the free flow of the smoke and waste gases. I used to be a bit of an ornithologist and liked to watch bird behavior, particularly in spring when their plumage was at its best. Spring is also the nesting season, and I was horrified to discover that birds sometimes nested in and around chimneys. Apart from the danger to the birds, this action can restrict the escape of the flue gases from the home. The outcome can be an unpleasant, smoky, or acrid atmosphere, with less air drawn into the fire — a further cause of inadequate combustion. For the same reason, flues should be constructed to be as straight as possible.

Even natural gas flues need an annual inspection, though the reason has more to do with deterioration than immediate hazards. Nonetheless, burners that are not serviced regularly can produce a lot of soot, and who knows what creatures are attempting to make *your* chimney into *their* home!

Much of this is common sense but was not general practice until com-paratively recently. My family was perhaps somewhat more enlightened than most. I well remember Grandpa Walker's little home in the late 1940s, when the chimney sweep came every springtime. I was given the job of going into the garden to watch for the brushes emerging from the top of the chimneys. I then had to run in and tell the sweep. (I'm sure he only wanted me out of the way while he did the job.)

Leakage

The gas explosion that led to Grandpa Walker converting back to coal was caused by leakage. Even in his day, this was extremely rare, due to the careful design of gas equipment and the tight regulations concerning its installation. These regulations have been tightened even more, both in the US and the UK. Nonetheless, a regular maintenance contract, a carbon monoxide detector, and knowing how to deal with a leak are essential even for today's more sophisticated consumer.

Leaking gas can be detected by its smell. When smelled, any flames or sparks must be avoided, the gas dispersed and the cause of the leak traced and dealt with. Gas companies have 24 hour a day, 365 day a year free emergency service, and they would rather turn out for a hundred (genuine) false alarms than one explosion!

On the morning of March 18, 1967, the oil tanker *Torrey Canyon* ran aground off Lands End in England—the first of a number of major tanker accidents around the world. Its entire cargo, approximately 860,000 barrels of oil, was released into the sea during the next twelve days. The Royal Navy finally bombed the tanker on March 28 to sink it and burn off the oil. The cleanup operation took years, but much was learned about how to deal with oil spills that helped in cleaning up later disasters, including the *Exxon-Valdez* spill.

Heating oil is more akin to diesel fuel than to crude oil, but is still unpleasant, smelly, and as difficult to remove from carpets as from the sea and nearby coastlines. A leakage in the house can be a costly business. Oil is also a fire hazard, though more difficult to ignite than gas. The defense is, once again, careful design, regular maintenance, and sensible use of equipment. Should a leak occur, action must be taken to stem it and repair the fault.

Basically, it is a matter of tracing the leak and closing the appropriate valves. If a fuel lift pump is installed, this should be switched off, too. Repairing a leak is usually beyond the ordinary homeowner's abilities, so send for your maintenance contractor and leave it to them. Your main responsibility will be clearing up the mess. Storage tank leakage is, as mentioned earlier, a different issue, with serious and expensive hazards to the environment. Rust perforation is the main worry, but leaking valves can also occur. So have your tank checked over from time to time—by the time a leak gets serious the cleanup alone will cost you dearly.

Solid fuel does not leak, but sparks may fly from log fires, and hot ash could fall onto the carpet. While these are unlikely to burst into flame, the carpet can still be seriously charred and ruined. My mom and dad rarely argued, but there was a regular bust up every few weeks over the use of fireguards. When Dad came home from work, he still wore his oily overalls and often flopped down on the floor in front of our coal fire. If he'd had a particularly hard day he would fall asleep for an hour or so. Trouble was, he sometimes forgot to put the fireguard up, and Mom worried about sparks hitting an oily patch on his clothes. This possibility "sparked" a row. Fireguards are useless anywhere except in front of the fire.

Electrical Heating

Electricity neither "leaks" nor generates any by-products (at the point of use). Any hazards come from either faulty equipment or misuse of equipment. In both cases, fusing and grounding provide effective fail-safe mechanisms, both of which operate by cutting off the equipment from the supply. Sparking can also occur, though its presence is easily detected. In all cases the appropriate action is to remove the equipment from the supply and either discard it or arrange a repair.

No flue or chimney is needed for a wholly electric system, but fireguards are advisable around an electric fire—even though there is no naked flame. The hazard is of flammable materials coming into contact with the hot wires and igniting. Long, flowing nightdresses are a particular worry. In the UK, these must be flameproof by law.

Convenience

Grandma and Grandpa were right about convenience. Carrying in buckets of coal and then carrying out ash is no fun when you are in your 70s—to say nothing of getting up on freezing cold days to make the fire. It must have been particularly tough on servants in stately homes who had so many fires to start before the lord and lady got out of bed. Then there's the dust, the need to use valuable space as coal storage, and so on.

Like many young children the world over, one of my favorite games was hide and seek—and one of my best hiding places was behind the coalhouse door. Our coalhouse was built with its door facing the back door of the house, about four yards away, so Dad had only a small distance to fetch the coal. Better-off people had coal cellars inside the house so they didn't have to go out in bad weather. I often got into awful trouble for hiding in the coalhouse, because inevitably I emerged covered in coal dust. I suppose this would not be a problem with wood storage, but a large space still needs to be set aside for it.

Natural gas and electricity are distributed directly to the premises where they are used and so do not need storage facilities. The situation is different with liquified petroleum gas (LPG). Individual large tanks, filled from time to time by the oil or gas company, can be provided, though premises are sometimes supplied from a central neighborhood tank with individual metering. A further alternative is to use exchangeable "bottles" of LPG. These tend to be more popular with the boating or trailer community, but there is no reason (other than price) why they shouldn't be used in permanent dwellings—though this reintroduces the inconvenience of having to physically handle the fuel.

Grandpa was a miner. Dad did slightly better for himself and became a maintenance engineer. The job of his team was to keep all of the equipment running properly in the chemical factory where he worked. He used to love it when the factory manager regularly reminded him: "Now Ernie, don't go fixing anything that ain't broke, will you?" Pa would touch his cap and put on his most serious expression to say: "You can be sure of that Mr. Webster, there's no time to spare for anything like that." He would then laugh all the way home to tell Mom the story. You see, he knew full well that without preventative maintenance, he would always be in a job, because things would always keep breaking. Furthermore, he would always have plenty of overtime (at time-and-a-half) because repairs take priority over everything else when production stops. In fact, he and his team became expert at making jobs last until the end of the working day, and sometimes into the weekend!

The same principles apply to the whole of a household heating system. If you want to avoid the inconvenience of breakdowns, periodic servicing and maintenance is essential. The nature of the fuel determines just how much and how often service is needed, but an annual safety check is advisable with all fuels. Preventative maintenance can be done at the same time. I strongly recommend maintenance contracts—the little extra they cost saves time and money in the long run.

On a day-to-day basis, the level of convenience of a system depends on the level of automatic control it has. In the fifties, our hot water heater had a thermostat so there was no problem of temperature control. Yet Mom and Dad regularly had arguments on bath night about who had forgotten to switch it on! This source of irritation was eventually resolved by also fitting a timer. Nowadays even better systems are available that make control possible over both time and temperature for the heating and hot water systems separately. Control systems are covered in full in Chapter 7.

The reliability of fuel supply is rarely a problem nowadays, and supply contracts from some companies are written to enable compensation to be paid to the consumer should it fail. Insurance can also be taken out to deal with this possibility.

Saving the Planet

When I first started lecturing back in the early 1970s, I taught a course called "Man and the Biophysical Environment," or MATBE for short. There were barely enough students to justify its existence, and most of those were regarded by many as the oddballs of the campus. It's very different today. Government environmental regulations seem to be under continuous scrutiny. Large sums of money are placed into research, and environmental policies spawn heaps of advice and exhortation directed toward ordinary citizens.

Yet there is no doubt that our individual efforts are important. My recycling of aluminum drink cans doesn't make much difference, but if we all recycle we can save the ore *and* the massive amount of energy used to process it. An apt slogan from Friends of the Earth, founded in the US in 1969 and now the largest environmental organization in the world, is "Think globally, act locally." It is important that we both do our bit *and* have our say.

Probably the greatest environmental advance in my lifetime has been in improved air quality. Soon after we first got a TV, my favorite program became *Dixon of Dock Green*. Dixon was a friendly London policeman who was admired and respected by all in his area, Dock Green. Weekly episodes told of his dealings with both criminals and the general public. The story writers at the time were aided by the phenomenon of the London smog, a thick mixture of fog and smoke into which people could disappear and reappear quickly at appropriate moments.

This smog was caused by the particulates released from burning coal. At that time, coal fires heated nearly all households, and many industries relied on coal-fired boilers. Stiff clean-air legislation followed, and a whole new industry for the manufacture of smokeless fuel began. A problem solved, or is it? Smog remains a problem in many of the world's great cities, including Los Angeles, Mexico City, and Bangkok, for example. The culprit these days, however, is road transport. Diesel fumes in particular, but also gasoline, contain particulates and chemicals that create the same problem. I am left wondering how long transport powered in this way will remain viable.

What of other fuels and environmental impacts? We confine ourselves here to brief comments, but I include books in the Further Reading section that will enable you to follow up in more detail if you wish.

Natural gas releases carbon dioxide into the atmosphere when it is burned, but not as much as coal for each unit of energy released. It is itself a potent greenhouse gas, so leaks in the system are also an issue. On the other hand, it has low emissions of toxic gases and particulates. Burning residential heating oil has an impact similar to natural gas.

The renewability of wood is a strong plus environmentally, though it emits carbon dioxide and other potentially harmful gases when burned. Of the other renewables, small-scale hydroelectric generators are available for insertion in a nearby river or stream, as are small-scale wind generators, but continuity of supply is a problem (the wind doesn't always blow!). The heat from the sun—the solar gain—is a free source of renewable energy, though just how useful it is depends on your particular climate. More details of renewable options are provided in Chapter 8. For now though, you could choose to pay a bit more to get electricity from a company that encourages generation from renewable sources. So, why not see what is available in your locality?

Decisions, Decisions

If you are in the fortunate position of being able to choose the fuel to have in your home, making a decision can seem overwhelming because of the number of factors involved. I have tried to help in table 4.5, where I have ranked fuels according to each criteria—the more stars the better. Remember, though, that this is subjective, so others may disagree with some of my ratings. Box 4.3 gives some tips on fuel safety, which I hope you will find useful.

Table 4.5   A Simplified Comparison of Fuels

	Electricity	Natural gas	Oil	Solid fuel	Renewables
Availability (now)	★★★★★	★★★★★	★★★★★	★★	★★
Availability (future)	★★★★★	★★	★★	★★★★	★★★★★
Versatility	★★★★★	★★★	★★★	★	★★★★
Cost	★★★	★★★★★	★★★	★★★★★	★★★
Safety	★★★★	★★★★	★★★★★	★★★★	★★★★
Convenience	★★★★★	★★★★	★★★★	★★	★★★
Environmental impact	★★★	★★★	★★	★★★	★★★★★
TOTALS	30	26	24	21	26

Box 4.3. Top Safety Tips for Fuels

It is difficult to give general tips, because choice of fuel depends on availability and individual preference. Consequently, I confine myself here to safety tips.

- Set up a program of safety inspections of your home (a) by a professional annually and (b) by yourself quarterly or near the beginning of each season.
- As you check around your home, identify hazards and decide what preventative measures are needed.
- Decide what you would do if a risk became a reality.
- Ensure that you know where all safety valves and switches are located and check that they can be easily closed and opened.
- Ensure that all grids, grills, and registers are clear of obstructions.
- Ensure that all fires (including electric ones) have a fireguard.
- Make sure that you have readily available the emergency telephone numbers for all of your utility companies.

By far the greatest hazards arise from a gas leak. The same procedure applies to natural, propane, or liquefied petroleum gas. If you suspect a leak in your home, carry out the following actions *in this order:*

1. Put out any flames, cigarettes, etc. Do *NOT* operate *ANY* electrical switches or controls, because of the danger of sparking.
2. Evacuate the premises, leaving door and windows open to disperse the gas.
3. Call for help, using a cell phone or a neighbor's phone.
4. Switch off the gas supply to the premises if the valve is easily accessible and it is safe to do so.
5. Alert neighbors.

Gas companies operate an emergency service 24/7 every day of the year. There is generally no charge for such an emergency. For suspected outdoor leaks, evacuate the area and call for help as above.

CHAPTER 5

Flickering Flames

The Science of Fires and Fireplaces

Making the Fire

Imagine yourself as a child, lying in a warm comfortable bed, slowly wakening on a cold frosty morning. Listen. What sounds do you hear? The howling wind? Birds singing? Traffic rumbling by, or perhaps street sounds as people make their way hither and thither? For me, it was none of these, for I invariably woke to the noises of thumping and banging, crackling and blowing, and occasionally loud cussing, all coming from our downstairs parlor. Dad was making the fire.

Making a fire in a cold climate is a skill learned early in life, but somehow Dad never quite got the hang of it. The process began with "chopping the sticks." A sack of wood scraps was bought from the local joiner's, and on a dry day Pop would carefully chop them into short thin sticks with his hatchet. They were then stored in the coalhouse, a large shed opposite the kitchen door.

Every morning he took the previous day's ashes to the trashcan and went to the coalhouse to fetch some coal and sticks for the parlor fire. There was a grate in the fireplace, and first he crumpled up yesterday's newspaper to make

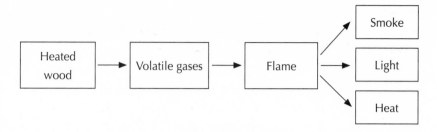

FIGURE 5.1.   Lighting the fire

a large ball. The sticks were laid upright on the paper and surrounded by coal. He would then light the paper with a match and wait. On a good day the paper would light the sticks, the sticks would light the coal, and the resulting fire would warm the room.

Often, however, things went wrong. If the fire didn't "catch," he blew on it from below to supply more oxygen. If he blew too hard, he would blow out the small flames and have to start again. If he didn't blow hard enough, it didn't make enough difference and again the fire went out. Sometimes he would "draw" the fire. This involved holding a large sheet of newspaper across the fireplace, leaving a small gap at the bottom. Warm air rising up the chimney would be replaced by cold air directly onto the coals, supplying much more oxygen than is possible by blowing. However, the newspaper sometimes caught fire, leading to lots of loud cussing, a room full of smoke, and burned hands.

This little bit of family history illustrates the basic requirements for lighting a fire. A fuel is needed, which must be ignited with either a flame or a spark. For solid fuels, dry tinder (Dad's paper and the sticks) is also necessary. Finally, there must be sufficient airflow to the fire and a safe escape route for the waste gases.

As any firefighter will tell you, the science of fire is not quite as simple as it first appears, for it can behave in complex and unexpected ways. A good starting point for understanding it is shown in figure 5.1, which shows the sequence of events for igniting a wood fire. The sequence is similar for a coal fire.

A spark or flame ignites the tinder, which then heats the wood to a temperature of around 300°F, sufficient to decompose it into charcoal, ash, and volatile gases. The temperature continues to rise, and when it reaches 500°F, the gases burn to form water, carbon dioxide, and various other waste products, which generally disappear harmlessly up the chimney as smoke.

If you watch this process carefully, you will initially see unburned smoky gases rise from the surface of the wood where the temperature isn't yet sufficient for it to ignite. As mentioned in Chapter 4, these unburned gases can condense to form thick, gooey creosote inside the chimney, creating a dangerous fire hazard. Anyone with a wood or coal fire is therefore strongly advised to have an insulating liner fitted inside the chimney to prevent this condensation occurring. As you continue to watch, flames appear that are hot unburned carbon particles being carried upward by rising air. In the absence of a chimney liner, these too may be deposited inside the chimney as soot—another potential hazard.

After a while, when all the volatile gases have been driven off, you will see the warm red glow of charcoal as it slowly burns and releases radiant heat into the room. The whole process is self-perpetuating: the heat energy released from burning keeps it going until only ash remains.

Why, then, did Dad have so much trouble? Getting the air supply to the fuel right is only one of a number of potential problems. The size of the fuel is also crucial. If it is too big, then massive amounts of heat energy are needed to raise the temperature enough to allow ignition to occur (try lighting a log). The size of Dad's sticks or his coals could simply have been too large. Dampness is another crucial factor. The water in damp wood has to be driven off before the fire can ignite, needing large amounts of energy. "Keep your tinder dry" is a survival tip from eighteenth-century pioneers, but our coalhouse roof wasn't completely waterproof, and Dad sometimes forgot to find a dry spot for his sticks. I reckon this must have been the cause of much of his cussing!

Ignition of the tinder and then of the fuel can thus be very tricky, so much so that the firekeeper has a very important role in Native North American ceremonies. In the past, they were responsible for maintaining and keeping the fire alight for many weeks. Perhaps the most well-known European equivalent was the cult of the Roman God Vesta, in which four vestal virgins tended the sacred fire in the temple. The loss of the flame was thought to be such a misfortune that severe punishment, usually the death of the vestal virgins, followed.

Fireplaces

Although they are no longer essential, fireplaces remain an important part of many of our homes. Their flickering flames have a quality that creates an appealing atmosphere of tranquility, and the hearth remains our usual family gathering place. It is there in front of a warm fire that we chat, plan, and celebrate. So important to us is the peace and calming effect of the flickering flames that modern gas, oil, and electric fireplaces are often designed to mimic burning logs or coals so that we can enjoy their unique ambience whatever fuel we use.

There is no great mystery about the source of the flickering, no matter what you believe about the psychological effects. Neither logs nor coal are homogenous materials, and consequently there will be small random changes in the fuel concentration at the points where the flame meets them. This is what gives rise to the flickering. Effectively, the air/fuel ratio is varying randomly with time, leading to periods of high flame followed by periods of low flame. The same effect can also occur because of small air movements in the vicinity of the fire.

The modern fireplace has its origin in medieval times. Figure 5.2 shows the fireplace in the kitchen of Hampton Court Palace, the home of King Henry VIII. The sheer size of the fireplace is impressive, but notice the burn marks above it—evidence of what happens if you get fire-making seriously wrong! Wood, from the many trees needed to keep the royal entourage warm and fed, is stacked forward and to the side of the fireplace, but away from the fire itself.

The fire is built in a metal basket or grate, so that warm waste gases escape up the chimney and draw more air through the fire, keeping it supplied with oxygen. Early fires were built directly on the ground and were prone to going out through lack of oxygen. Building a fire on a grate above ground level was therefore a major technological advance. Energy loss up the chimney remained a major issue, however.

The energy emitted is largely radiant heat from the red-hot fuel, and consequently the warming effect decreases noticeably with distance from the fire. Also, the rising waste gases draw more air in from the room, leading to cold drafts across the floor. The ash has to be removed frequently, and the fire kept

FIGURE 5.2.   A Tudor kitchen fireplace

supplied with fuel. This is fine if you have a small army of servants to look after you, but jolly inconvenient for the rest of us!

Henry VIII's chimney is brick lined and the flue tapers upwards. Various fire management and cooking tools can also be seen on the left of the fireplace. What is noticeably missing is any kind of fireguard. When I first saw it, I was left wondering how many tipsy cooks fell into the fire, or how many servants were injured by flying sparks.

Fireplaces in the homes of both nobles and commoners were smaller versions of the one at Hampton Court and remained much the same in design until the end of the eighteenth century, when Count Rumford suggested considerable improvements. Count Rumford was born in Woburn, Massachusetts, and was known then as Benjamin Thompson. He remained loyal to the Crown and had to flee to Europe, where he eventually became a "Count of the Holy Roman Empire."

Count Rumford made fireplaces much smaller, with the fire further for-

FIGURE 5.3.   A Rumford-style fireplace

ward and thus displayed at a much wider angle. This allowed heat to radiate better into the room. He also rounded off the throat of the fireplace so that gases could escape more easily through the flue, which was built *into* an outside wall. The outcome was a more efficient fireplace with a pleasing classical elegance.

Figure 5.3 shows a Rumford-style fireplace at Willersley Castle in Derbyshire, UK. This was built in the late eighteenth century for Sir Richard Arkwright, the "father" of the factory system, who made his money from designing, building, and operating the world's first successful water-powered spinning mills. Interestingly, the first American factories (in New England) used Arkwright's mills as their model, and his UK mills have become a World Heritage Site.

Many older homes still use Rumford fireplaces, and from time to time they enjoy a fashionable revival as a feature of new homes. But just how efficient are they? The answer is an unimpressive 10%–15%, meaning that mas-

FIGURE 5.4.   Tudor decorative chimney (Hampton Court)

sive amounts of fuel are required to produce limited amounts of heat energy. Wood may be a renewable resource, but if it were the major fuel for our homes, we would not be able to grow timber nearly fast enough to replace the amount we burn. Fortunately, today much higher efficiency can be obtained, almost entirely due to better flue and chimney design. But solid fuel fireplaces remain comparatively inefficient compared with other fuels.

Venting-Flues and Chimneys

Early decorative chimneys like those serving Hampton Court (figure 5.4), provided little more than an opening to the atmosphere of sufficient height to draw air through the fire below. However, the flue often took several twists and turns before reaching the chimney outlet, thereby allowing soot to be trapped in the bends and making it difficult for the smoke to escape. The result was poor indoor air quality.

Improvements took place over the years, one being the *throat restrictor* (an adjustable flap) placed in the chimney, which allowed further control of air to the fire by restricting the escape of hot air through the flue. The fire could therefore be "damped down" once it was in full operation. A further refinement was to fit a metal hood above the fire and connect it to the chimney. This was heated by the escaping warm air and provided some convected heat to the room.

There is still another refinement to consider. I remember Mom letting out the most fearsome scream early one Sunday morning. I leapt out of bed and hurried downstairs half naked. Dad rushed in from the garden, and several neighbors appeared at their doors to see what the commotion was about. A bird had fallen down the chimney into the fire. A terrible death and not a pleasant sight. I didn't fancy any breakfast that morning. These days, chimneys have a chimney cap to prevent this kind of occurrence while still allowing waste gases to escape. It also prevents downdrafts in high winds, which can blow smoke into the living room or even extinguish a fire that has just been lit—another cause of Dad's early morning cussing.

The fluing and venting arrangements of our house were good in their day, but nowadays would be regarded as very inefficient and far from ideal. Much of the low efficiency arose from the waste combustion gases rushing up the chimney and effectively sucking warm air out of the room. This was then replaced by *cold* air, which had to be heated, meaning that warm air was constantly being "thrown away" and replaced. Our unlined brick chimney also lost heat energy to the outside by conduction through the walls.

When my wife and I moved into our current home, the heating system was far down on our list of priorities. We were more concerned, as most people are, with location, and for us it was perfect. A small Peakland village, surrounded by countryside yet within easy commuting distance of Manchester, was great. When I did get around to thinking about the heating, I was disappointed. The gas fire has a back boiler, meaning that the boiler for the hydronic central heating system is hidden away below and behind the fire. Both fire and boiler share the same vertical flue, exhausting the gases to the outside via a chimney on the roof. It is an inflexible arrangement. There is simply nowhere else we could fit a free-standing boiler in the whole house, short of building a boiler house onto it.

Still, the fireplace looks okay, with flickering flames of gas burning on the (ceramic) coals to give a warm romantic appearance. It is also easy to use and

FIGURE 5.5.   My current gas fire

provides adequate warmth to our parlor. The current arrangement is shown in figure 5.5. The drawback is its low efficiency, which does neither my wallet nor my environmental credentials much good. It is technically very like my boyhood coal fire, in that it uses already heated air from the room as combustion air. There also have to be grids to the outside within the room to ensure an adequate air supply to the burners (see figure 4.2.). This type of fireplace is known technically as a gas *B vent* system.

One day I hope to get enough money together to rip out the system and start again with a more efficient modern *direct-vent system*. In these, combustion air is provided from outside via a sealed intake duct, and a separately sealed flue removes the waste gases. Airtight doors are fitted to the fire to prevent any warm room air being drawn into the system. Finally, pyroceramic transparent windows are fitted in the doors to transmit infrared radiation into the room while al-

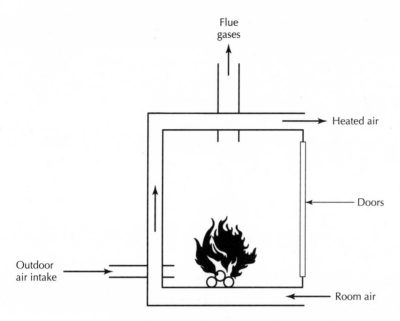

Flue
gases

Heated air

Doors

Outdoor
air intake

Room air

FIGURE 5.6. A modern direct-vent fire

lowing a good view of the fire. Thus, the calming effect of the flickering flames is not entirely lost, and indoor air quality is much improved. The efficiency of this type of fireplace lies in the range 50%–70%, a huge improvement on the 15% of the Rumford fireplace. The arrangement is shown in figure 5.6.

Ventless fireplaces are now on the market, but there are concerns about their safety. Not only do they use warm room air as combustion air, but they also exhaust directly into the room. Accurate and adequate sizing is therefore crucial to ensure that sufficient air reaches the burners and there is no dangerous buildup of exhaust gases. They could be useful as backup systems during power outages, but I would be very nervous about using them as my main heating source. They are illegal in some states and municipalities.

Because I was born into a family of miners and lived on a coalfield, a house without a chimney looks very strange indeed to me, yet there is no need these days to have one. An efficient alternative is to use two concentric pipes going to the fire, the inner one carrying away the waste gases and the outer one bringing in the combustion air from outside. These pipes can pass horizontally through an outer wall without the need of a chimney. This is called a *coaxial vent* in the US but is known as a *balanced flue* in the UK (another example of our being "separated by a common language"). They must be carefully positioned so that waste products can easily disperse and fresh air can enter. Local wind patterns may therefore render them unsuitable for some locations. They also must not allow waste gases to enter through nearby windows.

Practice is always more complicated than theory, and detailed type, sizing, and positioning requires expert advice. We have our fire checked annually by a qualified engineer, and he always checks the efficiency of the venting as well as the correct working of the fire. It is well worth the cost to know that everything is okay.

Stoves

When I was about 8 years old, we had improvements made to our little house involving, among other things, the installation of a gas *cooking* stove. The early morning cussing was now much decreased, for Dad would light all the burners and leave the oven door open to keep him warm while he was making the fire. Mom even caught him drying out his sticks in the oven on one occasion—the cussing was replaced by rowing that morning! The problem with Dad's approach is that cooking stoves are not designed for efficient space heating and are rarely vented to the outside. The level of carbon dioxide and water vapor in the kitchen therefore made it very muggy, particularly since we didn't have an extractor fan.

In using a cooking stove for space heating, Dad was treading a well-worn path. Stoves have been around for a surprisingly long time. The Chinese of the Qin Dynasty used clay cooking stoves fired by wood or charcoal at least two centuries before Christ. These gave out considerable heat and were effectively dual-purpose, as were open fires such as the one in the kitchens of Hampton Court. However, as time went by, the two functions became separated. These days a stove is only designed to fulfill one of these functions.

A stove is essentially an enclosed receptacle for a fire. For space heating, some heat radiates from its surfaces into the room and some is conducted through its surfaces to the air. It then rises by convection to fill the room with warm air. This arrangement improves air quality, reduces fire risk, and increases efficiency, compared with an open fire.

In Europe and America, stoves didn't get going in earnest until the eighteenth century with the invention of cast iron, an excellent conductor of heat that also radiates it well. Cast iron is therefore ideal for making stoves, with the added bonus that the use of decorative molds makes them more acceptable in living areas. Like most Brit kids, I used to love old cowboy movies. A feature of these was often the bunkhouse, in the center of which was invariably a cast-iron stove that had to be regularly "fed" with wood. In energy terms these were terribly inefficient, but they were nonetheless the starting point for developments leading to their highly efficient descendants of today.

The American contribution to stove technology is highly significant, and to this day the US remains the world leader. This contribution began in 1742, with a carefully designed wood-burning stove invented by Benjamin Franklin. To create the Franklin stove, he slowed down the flow of the hot waste gases by passing them through additional internal pipework. More heat energy from the gases flowed into the room via the pipework rather than escaping through the stovepipe to the outside.

Franklin stoves increased fuel efficiency to around 25%, much larger than fireplaces of the time, whose efficiency could be as low as 5%. Imagine what this meant in terms of the amount of wood that had to be cut to keep a family over a winter. Approximately the same amount of energy could be obtained from a fifth as much timber using a Franklin stove. These stoves are still available from specialty suppliers to give a retro look to a home.

Not everyone, however, welcomed stoves. The influential American essayist Henry D. Thoreau was one who didn't. Born in Concord, Massachusetts, he influenced Gandhi, Martin Luther King, Jr., and the British Labour movement through his essay on "Civil Disobedience" (1849). His best known work, however, was *Walden; or, Life in the Woods* (first published in 1854 and still in print), which described a two-year period in his life in which he withdrew to live in the woods. Consider this extract from Chapter 13, "House Warming":

It will soon be forgotten, in these days of stoves, that we used to roast potatoes in the ashes, after the Indian fashion. The stove not only took up room and scented the house, but it concealed the fire, and I felt as if I had lost a companion. You can always see a face in the fire. The laborer, looking into it at evening, pulifies his thoughts of the dross and earthiness, which they have accumulated during the day. But I could no longer sit and look into the fire.

Let us not underestimate the psychological effect of flickering flames.

Modern Wood Stoves

Until comparatively recently, wood has not been a convenient fuel, for it has to be cut, stored, and its ash removed. What's more, the fire often needs refueling several times a day, especially in the depths of winter. Add to this the fact that the calorific value of wood varies according to how dry it is and the particular species of tree from which it has been cut. Oak has a higher density than pine, for example, which gives it more material for burning per cubic foot. It is easy to see therefore why it has not become a common fuel.

Yet wood has undergone something of a comeback over the last ten years or so, partly as a result of rising environmental concern and partly because the inconvenience factor has been considerably reduced by the increased availability of wood pellets, which are supplied ready to use in twenty- or forty-pound stackable bags. These are made from sawdust, scraps, or chippings from lumberyards, furniture makers, and so on, to standards set by the Pellet Fuel Institute. Less wood waste therefore goes to landfill, with reduced environmental (and financial) costs.

The price and availability of wood for fuel varies, but the running costs can be competitive with other fuels, since modern advanced combustion wood-stoves can achieve efficiencies in the range of 65%–80%. They do this by having two combustion zones, compared with conventional designs, which only have one. The first is the ordinary wood fire, and the second is a zone above that burns off the volatile gases that would otherwise escape through the chimney. The heat energy in these gases is thereby reclaimed. Incorporating a direct vent arrangement further increases efficiency, as do arrangements for extracting energy from the flue gases before exhausting them to the outside. These gases meet pollutant emission standards set by the EPA in the United States and the CSA in Canada.

A further refinement allows the fuel to be loaded into a hopper, roughly once a day only, and delivered to the fire by turning a corkscrew-shaped auger.

Air is also delivered in a controlled way by an electrically operated fan. These stoves thus rely upon an electrical supply. The cost of this supply is relatively low, but an outage disables the appliance, so emergency backup is advised. More details may be obtained from www.hearth.com/what/specific.html.

Coal Fires and Stoves

When Dad decided against a coal fire, he chose a gas fire. Grandpa, however, changed his for a Belling-style electric fire, having tried a gas fire for a while but then rejecting it for fear of an explosion. Nonetheless, his decision was a bit strange. He was a coal miner for many years until his deteriorating health meant he had to have a surface job. As I mentioned earlier, he was still entitled to a generous annual allocation of free coal, and I'd have thought he would have persevered with his coal fire. Sure the inconvenience of it got to be more and more of a problem, but this wasn't his main reason for abandoning it. One day he explained it to me: "You see, Lad, I spent twenty years down t' pit. It wore dusty, dark, dangerous an' bloody 'ard work. I 'ad pals badly injured and one or two killed. An' we 'ad t' fight [strike] for every penny—they cut our wages an' then tried to starve us back to work. All to keep the bosses in luxury. I'm sick o' coal. It would suit me if I never saw another lump, let alone 'ave any in t' house." There's no answer to that is there?

Still, there remains a significant minority of people both in the US and the UK who rely on coal to heat their homes. We need say little here about coal fires and equipment, for the principles are exactly the same as for wood stoves. Detailed design, however, is different enough for the fires and stoves not to be dual fuel operable; you can't burn wood in a coal stove or vice versa. One advantage of coal over wood is that it has a higher heat content per ton (a higher calorific value), so refueling can take place less often. On the other hand, it is nonrenewable and therefore less acceptable environmentally.

Gas Fires and Heaters

Henry Thoreau's writings stirred many people, but my Dad would have seen him as a hopeless romantic, if not a crackpot. He would wonder, as I do, what laborers Thoreau was talking about when he wrote: "You can always see a face in the fire. The laborer, looking into it at evening, pulifies his thoughts of the dross and earthiness, which they have accumulated during the day." Dad was active in the Union movement, and I have memories of having to sit quietly in the kitchen with Mom while "the committee" met in the parlor to decide how to run the latest strike. Sure he looked up to intellectuals, and no one was more proud than he when I made it to university, but he simply wouldn't have understood a guy who gave up a good teaching job to go and live in the woods. As for looking into the fire to purify his thoughts, the laborers Dad and I knew would be insulted at the thought that they needed to do any such thing. Seeing faces in the fire would simply confirm that he was a "wrench short in his toolkit."

When we eventually "moved up in the world" and purchased a gas fire, Dad had no more getting up in the morning to make the fire, no more going out into the cold to bring in the coal, no more clearing the ashes, and no more worries about sparks flying into the room. He just switched it on. As an added bonus, it even cost less to run. What luxury!

Gas Fires

As long ago as 1626, French explorers found Native Americans igniting a mysterious gas around Lake Erie, the first known American use of natural gas. Its first commercial use was in 1816, when the streets of Baltimore were lit by gas manufactured from coal. Indeed, throughout the nineteenth century, gas was used almost exclusively as a source of light rather than heat.

The yellow flame used for lighting was, as discussed in Chapter 4, due to incomplete combustion arising from inadequate oxygen supply. The flame's source of oxygen is the surrounding air, which is swept into it and rises with it by convection. But not nearly enough air can be supplied in this way for complete combustion, which requires an optimum mixture of about one part gas to three parts air.

It wasn't until nearly seventy years after those natural gas streetlights that the German Robert Bunsen perfected his burner design that made gas heating a viable commercial proposition. His simple but brilliant idea was to mix the gas with air *before* it reached the flame, which produced the almost colorless flame with the blue cone characteristic of all gas burners. The temperature just above this cone can reach 2700°F, plenty hot enough for cooking and heating!

The design of burners for gas fires is still based on Bunsen's original ideas. Gas is introduced through a jet, and air is supplied through a throat-like opening within the pipe leading to the jets. This opening must be kept clear at all times to ensure that sufficient air can enter it. Control is automatic, for when the gas supply is turned up the gas rushes more quickly to the burner, drawing more air into itself, rather like the flood waters of a stream dragging parts of the bank with it. The whole thing is carefully engineered to ensure that all the gaps and jets are the right size, and there are often a number of small jets arranged to fit the shape of the appliance.

An early problem with gas fires and stoves was the arrangements for ignition. Long tapers or special gas lighters were needed to avoid burning the user. Later these were abandoned in favor of pilot lights, permanently burning small flames from which the main burners could be ignited when the gas was switched on. This, however, wastes gas, and nowadays press-button spark ignition is used: one small step for the environment.

The best-performing gas fires in terms of cost and energy efficiency are the enclosed direct-vent design, as shown in figure 5.6, where 100% of the combustion air is drawn from outside the house and 100% of the exhaust air is expelled to outside the house.

Gas Logs and Coals

The first gas fire that Dad bought for his family was efficient and convenient, but both Dad and I would have to admit that something was missing. It was unrealistic and had no flickering flames. Today we can have the best of both worlds, as many gas and electric fires are designed to mimic the effect of a log or coal fire (see figure 5.5 again for a burning coal effect).

For a burning log effect, gas logs are near perfect. They are made of a ceramic material (something like pottery) cast from real logs to give realistic details of the wood's knots and bark. They do not burn, although a layer of carbon eventually builds up on the surface, giving them a pleasant charred look. They can even crackle like a real log fire, all without its mess and inconvenience. The log is supported by a metal frame, and the gas burner beneath it provides adjustable yellow-orange flames, with a range of effects from small and flickering to large and blazing. The only maintenance required is to clean the logs with a soft brush from time to time—though an annual safety check of the whole fire by a qualified engineer is essential.

Gas logs simulate a wood fire very well, but you may recall that a yellow flame is a sign of incomplete combustion, with the attendant danger of carbon monoxide poisoning. Reassuringly, however, its detailed design makes a gas log fire at least as safe as a real wood fire (provided it has been approved, installed, and maintained to the appropriate standards).

Gas logs are sold in vented and vent-free types. The vented ones replicate a wood-burning fireplace and are very realistic, with a choice of different styles, types, and finishes. The downside is that, just like a wood fireplace, most of the heat goes up the chimney. Consequently, they are great for romantics, but poor for environmentalists. I wonder what Thoreau would have made of them! Vent-free logs are much more efficient (90% or more is achievable), and thus cheaper to use, but there are safety concerns. They must be fitted with an ODS (*oxygen depletion sensor*), which shuts off the flame if the oxygen in the air drops below a safe level, and safety issues remain for people with asthma or respiratory weaknesses. Some states (California, Massachusetts, and a few others) don't allow the use of vent-free logs. They are also not as realistic as vented gas logs.

Gas Room Heaters

Direct vent gas-fired room heaters have received a lot of attention in the last ten years or so. They look like large wall-mounted radiators, though most of their output is convected heat. Fitted beneath a window, they can be vented to the outside through a coaxial flue. They incorporate a gas burner, air filter, gas valves, a blower, and a thermostat within a painted metal cabinet.

The downside is the lack of a visible flame to warm the heart, but they can be very efficient and economical. Provided there is a gas supply nearby, installation costs are low, and even in the mild climates of California or Texas, installing one could be the smart thing to do. This is particularly true in a small home, where a properly sized and located heater could provide all the heat needed, especially if there is an open-plan layout with dining room and parlors interconnecting.

Room heaters can also be used effectively alongside a central heating system. Suppose you have a large home with an old system and you don't fancy replacing it. The *addition* of a gas heater could reduce both your bills and the overall energy used. There are two reasons for this; first, a modern heater will be more efficient than an old central heating system, saving money whenever it is used instead of the existing system. Second, a room heater tends to be used only when heat is required, whereas a central heating system is often left on to heat the whole house unnecessarily. The downside is that maintenance costs will be higher, and controlling two parallel systems automatically is difficult if not impossible. Personally, I would view this arrangement as an interim measure until I had the funds for a complete upgrade.

Ventless room heaters are also available, but suffer from the same problems as unvented gas fires. They are cheaper to buy but there are serious worries about their safety, and they are banned in some areas. Few people recommend them.

Gas Infrared Fires

When Dad abandoned coal and bought our first gas fire, it was of a different type, known as a gas infrared fire, which was much more common then than now. For these, the main heat sources are ceramic *radiants* which are heated to red heat by the gas flame. This gives a warm glow, with the flame just visible behind the radiants. They also provide some warm air to the room, giving them a relatively large output for their physical size. The lack of a visible flickering flame means they are not as nice looking as other gas fires and therefore they are not popular nowadays, though they are still available.

Oil Fires, Stoves, and Heaters

The first televised war, for Brits, was the first Gulf War of 1990–91. The images that stick in my mind are of the burning oil wells, with the air for miles around full of thick, black, acrid smoke, and blobs of floating oil around the edges of the Kuwaiti beaches. It seems incredible that the same stuff, after refining, can be safely burned in the fires and stoves of our homes.

The dark, smelly, thick crude oil is chemically a complex mixture of hydrocarbons, compounds consisting of carbon and hydrogen atoms only. Different hydrocarbons are formed by different arrangements and different numbers of these atoms. Mixed together like this they aren't much use, but they can be separated into a wide range of useful materials.

I haven't traveled much, but the little I have leads me to wonder if there is any country in the world that hasn't discovered how to make strong liquor. The Irish have poteen, the Americans moonshine, the Ugandans waragi, and the Scots whisky. The process involved in making all of these is distillation, and the stills needed are easy to build and conceal, so illegal backyard manufactured liquor is a problem for the authorities. Since the useful materials in crude oil all have different boiling points, they too can be separated, in a process technically known as *fractional distillation*.

The fuel oil used in domestic boilers is chemically very like diesel, and like all hydrocarbons burns in air to produce water and carbon dioxide according to the equation

$$\text{fuel} + \text{oxygen (from air)} = \text{heat energy} + \text{water} + \text{carbon dioxide.}$$

Oil burners within our domestic appliances, are pieces of precision engineering. In order to produce a steady hot flame, the oil must first be either vaporized or atomized. In a vaporizing burner, the oil is vaporized by contact with a heated surface and mixed with a controlled amount of air before being ignited electronically. Adjusting the amount of oil fed to the burner controls the output. This design is inherently reliable, as any engineer will tell you, simply because there are no moving parts to wear out or jam.

Part of the clutter in my garage consists of aerosol touch-up paint for cars long ago sold; after all, you never know when such things might prove useful! An atomizing burner operates in a similar way to an aerosol. In an aerosol, a liquid is vaporized by being blown through a small nozzle, either propelled by a gas from within the canister or by a knob on top of the can being manually pumped. In this kind of burner, oil is sprayed into the combustion chamber, propelled by a blower and ignited by an electric spark.

Soon after we first had a diesel car, I got a panicked phone call from my wife, Irene, who had forgotten and filled up the tank with regular gas. Fortunately the engine was undamaged, but the fuel system had to be drained and cleaned out at considerable cost. It is similarly important that the correct grade of fuel is used in oil burners. These are graded 1 to 6 by the American Society for Testing Materials (ASTM) according to their viscosity and sulfur content. Grade 1 is particularly adapted for vaporizing burners and Grade 2 for atomizing burners. Make sure you get the right one!

However, things are never as simple as they first appear, and so it is with oil fires and heaters. For example, the equation I gave for burning holds only under perfect conditions with just the right amounts of air and fuel in the burning mixture. As I write this, Britain's firefighters are tackling the biggest industrial fire in peacetime Europe at a fuel depot in Hertfordshire. Twenty storage tanks holding sixty million gallons of diesel, kerosene, and gasoline are ablaze, and the resulting black smoke plume covers a large part of southern England. The plume is mainly soot (carbon particles), indicating that the fuels have not burned completely, and this event illustrates what can happen when the air-fuel mixture is wrong.

In a domestic situation, a badly adjusted oil fire, heater, or furnace can also produce smoke, and soot can be deposited inside it. To make matters worse, sulfur occurs as an impurity in the heating oil. When this burns, it forms sulfur dioxide and sulfur trioxide, both of which dissolve in water to form a strong acid solution. A lined flue is required to prevent the acid from attacking the surrounding brickwork. Much like soot, solid yellow sulfur can also be deposited in the chimney.

All of this makes annual cleaning and maintenance absolutely essential. At this time, new nozzles should be fitted and adjusted, soot debris should be removed, and the pump motor should be checked over. Emissions tests are also required, making the process beyond the competence of the normal homeowner. An annual maintenance contract is advisable. Since the process is much more complicated, the cost of this will exceed that of a natural gas or LPG contract. If you are not connected to a central metered oil tank, then you will also need to have yours inspected from time to time to ensure that it continues to meet the appropriate safety regulations.

Oil fires and stoves are aesthetically much the same as gas models, the actual flame pattern varying from manufacturer to manufacturer. The choice between them is largely a matter of cost and availability.

Portable Kerosene Heaters

Beautiful though the Peak District of Derbyshire is, one downside is that we are prone to power outages in winter. So, one of our first purchases upon moving here nearly thirty years ago was a portable kerosene heater. The attraction is that they provide localized or emergency heating for homes at a relatively low outlay, using a fuel that is readily available and easily stored. There are literally millions in use in the US.

Portable heaters consist of a wick, which is immersed in a reservoir of kerosene near the base of the heater. The capillary effect ensures that the kerosene rises up through the wick, just like tea rising up into a biscuit dunked into it. When the wick is lit, air rises within the heater, is heated, and enters the room through slots near the heater's top. Kerosene heaters are normally tall and thin and finished for a pleasant appearance.

Kerosene, like fuel oil, is a hydrocarbon, so combustion produces carbon dioxide and water vapor. Since the heaters are almost always unvented, the room soon becomes muggy and uncomfortable. There is also a considerable fire risk with a naked flame close to a flammable liquid, and the possibility of consumer misuse is particularly strong. Some consumers have used the wrong fuel, placed heaters near combustible material, situated them where they can easily be knocked over, or overfilled them.

In response to these hazards, new models have safety features such as battery-powered ignition, automatic extinguishing devices, leveling indicators, and lift-out fuel canisters. Nonetheless, many fire officials and other government agencies are seriously concerned about kerosene heaters, to the extent that building and fire codes may not permit them in residential buildings. The Consumer Product Safety Commission's publication "What you Should Know about Space Heaters" (available from www.cpsc.gov/cpscpub/pubs/463 .html) gives authoritative advice.

Oh, and incidentally, we have never needed to use ours. The house, being well insulated, stays warm for several hours, and this has been sufficient to see us through power outages.

Electric Fires and Heaters

Electric fires, heaters, and heating systems abound. Technologically, they all rely on resistive heating. In my mid-20s, with an established track record as a teacher, I applied to join the Education Corps of the Royal Air Force. I was put through a process of officer selection, which included a mini assault course. For part of this I had to crawl through some narrow pipes. Wiggling my way through a tiny ill-ventilated space, against the clock, was no fun, and I got very hot. Much of my physical and emotional energy had become heat energy. I managed to push through, but to no avail. I failed overall, and the RAF decided they didn't need my services. (I was glad!)

Electricity passing through wires is a bit like that course. As the electrons go through long narrow wires—or simply wires made of a tough material—their energy becomes heat energy because the wires resist the flow. Hence the term "resistive heating." Producing useful electric heaters is therefore a matter of having the right wire (the heating element) and the right amount of "push" (the voltage) behind the electrons. There are a number of ways heating elements can be used to create convenient and effective heaters at the standard mains voltage of 110V.

The first feasible electric fires were shown at the Electric Fair of Crystal Palace in London in 1891. These were offered for public sale in 1893 by R. E. B. Crompton, and consisted of iron heating wires embedded in a layer of enamel, fitted to cast-iron panels. Unfortunately they kept breaking, because the iron expanded more than the enamel and frequently snapped. By 1904 a fire in which the heating elements were large sausage-shaped electric light bulbs (with carbon filaments), was on sale. The outsides of the tubes were colored, so they gave an orange glow and looked warm and welcoming. Alas, they gave out little heat. The carbon filament was also very fragile.

A robust heating wire was needed, one that would be efficient at converting electrical energy to heat energy and able to withstand high temperatures without oxidizing or losing its strength. In 1906, A. L. Marsh in England patented just such a wire, an alloy of a nickel and chromium wire known as *nichrome*.

By 1913 the Belling fire was introduced, and it is still available today with little change. The heating element is a nichrome wire wrapped around a nonconducting ceramic rod to give it strength. The element glows a warm red-orange color, emitting lots of radiant heat. It is mounted in front of a curved metallic mirror to direct the heat forward, in much the same way that the curved mirror of a flashlight directs light forward. Mounting the element inside a quartz tube is a further refinement. Quartz transmits the radiant heat but is electrically insulating, so should anyone accidentally touch it, they won't be electrocuted. (But don't try it—you may still get a nasty burn!) A one-kilowatt element is about eighteen inches long and enough to heat a small room.

An interesting American contribution was the development in 1926 of silicon carbide heating elements. They too have high resistance and glow when the current flows, but they are less stable and harder to manufacture than the nichrome wire design. By this time the wire element was well established in the US domestic market, and the silicon carbide elements weren't successful there. Nonetheless, the company is still in business producing special industrial elements; see www.globar.com.

The underlying technology of resistive heating is simpler than that for other fuels, so electric fires and heaters tend to be safer and more reliable. They also require much less maintenance. They are easy to control, with a switch that responds almost instantly, and thermostats and timers are easily fitted. Thermal cut-outs (also known as "kill switches") switch off the unit if it overheats, a useful additional safety feature. These convenience and safety factors have made them very popular.

Modern Electric Fires

Modern electric fires still use wire elements. What's more, they have increased in popularity in recent years as it has become more and more difficult to distinguish them from gas or wood fires. Technical advances have led to the incorporation of amazingly realistic flickering flames, so much so that some people like to have the flame effects switched on even when the hidden heating elements are off, just to enjoy the ambience they create. There is, of course, an extra cost to run these effects, but as a proportion of the total energy used this is slight.

Electric fires have a lot going for them in terms of consumer acceptability. Since there are no emissions, chimneys and flues are not needed, and there are no fumes, so indoor air quality is also good. Installation is easy and they are available in a tremendous range of sizes, finishes, and outputs. For many these advantages outweigh the extra running costs of an electric fire compared with a gas or wood equivalent. There are, however, two major problems with electrical heating.

In 1973, here in the UK, miners' unrest led to regular power outages and the imposition of a three-day workweek on most industries and services. Grandpa had retired by then, but Uncle Derek was involved in the strike, and we as a family wished him well. We contributed to the fighting fund and rejoiced when the government called a general election and lost. This is when we as a nation learned the foolishness of an all-electric home. Many people went to bed cold after a supper of sandwiches at that time. Outages continue to happen, rarely but also unpredictably. Consequently, non-electric backup systems are now regarded as essential.

The other problem with electric heating is its effect on the environment. Thermal power stations (the majority) unavoidably lose considerable amounts of energy in the generation process and emit large amounts of carbon dioxide, now known to be a factor in global warming. Not a good choice if you want to save the planet—unless you can locate a supply from renewable resources. This possibility is discussed later in Chapter 8.

Electric Heaters

There are many kinds of electric heaters, the simplest being the *panel heater*. These are wall mounted and look something like a thin radiator. However, they emit convected rather than radiant heat, for the internal heating elements simply warm the air in their immediate vicinity. The air rises and leaves the unit via grills at the top, while at the same time drawing more air in from the bottom. Although no radiant heat is supplied, the circulation of warm air generally heats the room rapidly. They are usually fitted with both time and temperature controls. Since they have no moving parts, they are silent in operation and very reliable. I have one in my conservatory (figure 5.7), which was an addition to our house. It would have been prohibitively expensive to connect the conservatory to the house's hydronic gas central heating system.

Fan Heaters

These can be thought of as simply a panel heater with an internal fan. They became available in the late 1950s, for surprisingly, it was not until 1953 that the *tangential fan* was invented (by Bruno Eck of Cologne). This is a long thin fan that can be mounted to run the length of a heating element. As with panel heaters, cold air enters through grills at the side or back of the appliance and emerges as warm air through separate grills at the front. A further refinement is that a choice of fan speeds can be supplied, and the fan can be operated independently of the heating elements, allowing them to blow out cold air in

FIGURE 5.7.   Electric panel heater

hot weather. As with panel heaters they can be thermostatically controlled, and equipped with a kill switch (i.e., a thermal cut-out).

Equally useful are portable fan heaters, which are simply plugged into the electricity supply. They are a valuable backup for, or supplement to, the main heating system. Mine are the sole heat supply for my garage, which is not connected to the main hydronic heating system. They would "save the day" if ever we had an outage in our gas supply. There are safety hazards associated with them, of course, and the Consumer Product Safety Commission's fact sheet gives an easy account of the various do's and don'ts (www.cpsc.gov/cpscpub/pubs/098.html).

A variation on this theme is the system I found in my hotel bedroom on a recent cultural visit to Stratford-on-Avon. Time and temperature controls were located near the door, and the unit blew out warm air on demand through the grid in the ceiling. At first I wondered if it was part of a warm air system or maybe an AC unit. I had a sneaky look one night (the grid was only held in

FIGURE 5.8.   Concealed fan heater

place by small crosshead screws) to find that it was simply a large fan heater, fixed just above the ceiling. I replaced the screws and photographed it (figure 5.8), and no one was the wiser!

Oil-Filled Radiators

In this form of electrical heater, the heating elements are enclosed inside a sealed steel casing and surrounded by oil. Heat energy flows from the elements to the oil, which rises and circulates throughout the radiator. Once up to temperature, the oil gradually releases the warmth to the room, even while the internal elements are off. The energy emitted is a comfortable mixture of radiant and convected heat.

Their main attraction is that the absence of a fan means that they are very quiet. They never need refilling or topping off, and all but the smallest models are mounted on wheels so that they can be easily moved around. A further benefit is that they are extremely safe, because the elements are sealed within the unit leaving no exposed parts. Their large surface area also means that they can give a good output without the surface temperature being so high as to burn anyone casually touching them.

THE CHOICE OF A FIRE, STOVE, OR ROOM HEATER involves financial, aesthetic, environmental, and convenience factors. Add to this the myriad of different models from a multitude of manufacturers, and the consumer can be forgiven for being confused. If you are in this situation, I hope you find the following summary table (table 5.1) and the tips in box 5.1 useful in making your choice. Then, I suggest you trot off to your local supplier to check availability within the category you have chosen. Good luck!

Table 5.1   Choosing Your Fires, Stoves, and Heaters

Category	Pros	Cons
Gas direct vent	Maintains indoor air quality Does not need chimney All combustion air comes from outside the home All exhaust air expelled out of the home	No direct access to the flame Carbon dioxide emissions, non-renewable fuel
Gas B vent	Aesthetically pleasing	Uses already heated air for combustion Requires permanent fixed opening to outside for adequate room ventilation Carbon dioxide emissions, non-renewable fuel

Table 5.1   Choosing Your Fires, Stoves, and Heaters, *continued*

Category	Pros	Cons
Gas unvented	Simpler and generally more flexible installation Low cost	Illegal in some states and cities Air quality and health issues Releases water vapor into room, which becomes muggy with possible mildew problems long term Requires permanent fixed opening to outside for adequate room ventilation Carbon dioxide emissions, nonrenewable fuel
Oil	Much the same as gas	Much the same as gas
Wood and coal	Aesthetically oustanding in terms of smell, glow, and crackle Firewood is readily and cheaply available in some areas. Wood is a renewable resource	Cleaning of the fireplace, removal of ashes, sooting of chimney May be difficult to light and needs continuous attention to keep it burning Unstable heat output Carbon dioxide emissions, coal is a nonrenewable fuel Poor indoor air quality is likely
Wood pellets	Clean burning with less frequent refueling Displaces fossil fuel use, saves landfill, renewable	Complexity of design entails more service and maintenance Carbon dioxide emissions
Electric	Ease of use: no cleaning, refueling, or igniting Enormous variety of choice	Susceptible to outages Tends to be expensive to run Inefficient overall (taking into account generation losses) Carbon dioxide emissions (from thermal power stations)

Box 5.1. Top Tips for Fires and Room Heaters

1. When installing heaters (other than electric ones), ensure you have either a good quality insulated flue liner or an adequate coaxial flue.

2. If you have a chimney, install a chimney cap to prevent downdrafts and things falling into your fireplace. Have your chimney swept and/or inspected regularly.

3. Check out the AFUE (annual fuel utilization efficiency) of prospective purchases.

4. Install only approved and accurately sized equipment for the job at hand.

5. Yellow and/or smoky flames indicate serious problems and are potentially dangerous, unless the equipment is specifically designed to give such an effect. Stop using the appliance and have it checked.

6. For gas appliances, ensure that the gas jets and air intake openings are not blocked or restricted.

7. Do not use portable kerosene heaters unless they are fully approved as safe. Use them strictly according to manufacturers' instructions and as backup devices only.

Ensure all your equipment is installed and maintained by a qualified engineer.

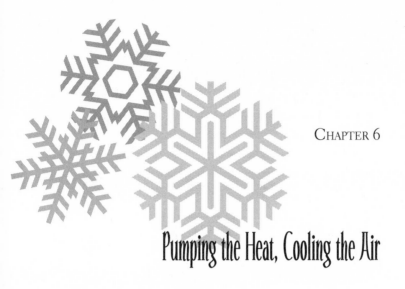

CHAPTER 6

Pumping the Heat, Cooling the Air

The Influence of Rome

Can you remember your first day of high school? I certainly remember mine. It's one of those turning points in life that leaves a great impression on us all. We were put into classes and learned pretty much everything together for the next few years. We were also put into *houses*.

A house, in the UK educational system, isn't a physical building but a group of students to whom you owe allegiance. In charge of each house was a house-master, one of whose roles was to encourage competition, especially in sports. You will be familiar with this kind of thing if you have ever read the novel *Tom Brown's Schooldays* (1857), by Thomas Hughes—a social reformer and founder of Rugby, Tennessee. The novel is particularly well known for the bullying of Tom by an older boy called Flashman, who "roasted" Tom in front of a fire. Tom eventually triumphs, and the book is an intriguing piece of social history. The house structure of the school is evident throughout. My own school had four houses: Troy, Athens, Sparta, and Rome. I was a Roman.

From that point on all things Roman interested me. Our local museum had some Roman artifacts, some of which came from a nearby long, straight lane—originally a Roman road—along which I often walked to reach Grandpa's house. It wasn't long before I also found out that Roman villas were centrally heated. This was ironic, for my family and millions like us still huddled around a coal fire in winter, two thousand years later.

In a Roman villa, a mortar floor rested on pillars of tiles, raising it about two feet above the ground. A fire was built outside, at the bottom of an exterior wall. A draft then took hot air across the underside of the floor and up through the hollow walls to chimneys, which were located in the corners of the building. Generally this worked well, but it was extremely labor intensive, as it took constant attention to feed the fire and remove the ashes. Slaves did this work and, consequently, only the wealthy could afford it.

Brits, at the time, got by with fires in individual rooms and with braziers, which could be moved around to different parts of their property. By contrast with the Roman system, this was primitive indeed. The Romans had a true system, in that all of the parts functioned together as a whole to accomplish the goal of good all-around warmth. In ours, various bits and pieces had been brought together without much thought about how they operated collectively.

When Dad eventually decided we could afford central heating, my childhood mind thought that the boiler would be installed in the center of the house, which was somewhere under the stairs. In fact, it was hung on a wall inside the kitchen. I was thinking of centers of circles rather than central as in Central Park, which is quite some distance from the center of New York City. I realized later that a central heating system is simply one where the heat energy is obtained from a single source and distributed from there to the other parts of the building.

Dad was quite happy with the completed system. Not so Mom. In an effort to save money, he had installed a partial central heating system—the bathroom was not heated—whereas she had expected a full system covering all parts of the house. She only started to feel better when she realized that the neighbors had only gotten background central heating, which merely gave general background warmth and had to be supplemented by booster heaters when it was particularly cold. There is another option Dad could have taken to reduce the cost: selective central heating, in which the whole house can be heated, but not all parts at the same time. The owner selects which parts to heat at particular times; for example, downstairs during the day and the bedrooms in the late evening and early morning. We will never know how Mom would have responded to this. Not very well, I suspect!

Table 6.1   The Percentage of US Households Using Particular Heating Systems (2001)

System type	Percentage
Forced air	60
Hydronic (pumped hot water)	11
Heat pump	10
Other electrical	8
Room heaters	4
Other	7

Source: Adapted from www.eia.doe.gov/emeu/recs/recs2001_ hc/t3_4b.html

I hope that this chapter will help you to understand heating systems and to make an informed choice should you decide to install one or upgrade an existing system. And, of course, in many states the problem is how to keep cool rather than warm, so air-conditioning is also considered. Be sure to seek more specific advice from a reputable source once you have a general idea of what you want.

In table 6.1, the popularity of different forms of heating systems across the whole of the US is shown. Forced air is easily the most common, but remember that average figures may conceal wide variations. Hydronic (pumped hot water) systems vary from 0% in the southern states to over 30% in the Northeast. The ability of heat pumps to supply both warm and cool air to the same ducting system makes them particularly suited to the climate of southern states, where there is a 40% usage. As you would expect, there are strengths and weaknesses to all systems; we will therefore consider them all.

Blowing Hot and Cold

Forced Hot Air Systems

The phrase "blowing hot and cold" comes to us from a Bible verse (Revelation 3:15), in which the church in Laodicea is castigated for being lukewarm. However, the author is talking about a lack of enthusiasm, rather than a building that is not warm enough. The most common US central heating

FIGURE 6.1.   Early warm air heating system. Photo courtesy of Belper North Mill

system literally blows hot or cold air, gently and controllably, into each room to achieve comfortable warmth. Unlike the Roman system, this arrangement warms rooms rapidly, because the air is heated directly without the walls or floor of the building having to be heated first. The Roman system was also entirely dependent on natural drafts, and the wind blowing the wrong way could render it ineffective.

Yet it was many years before the Roman system was improved. A man named William Strutt is credited with inventing the ducted warm air system in 1804, when one was incorporated into his cotton-spinning mill at Belper, Derbyshire. A stove low in the basement produced a current of warm air, which rose through unlined brick ducts to every floor of the building. Louver doors controlled the output into the rooms. The mill is still intact and may be visited by the public, and figure 6.1 shows the recently discovered site of the stove and ducting.

This particular mill is also remarkable because it was built on a cast-iron frame to support its five stories, plus attic. Most importantly, it was pretty much fireproof—earlier wooden mills were often destroyed by fire. Weight-reduction techniques were also incorporated in this building described as the forerunner of modern skyscrapers. One man who served his apprenticeship there, Samuel Slater, later sought his fortune by emigrating to the newly independent United States, illegally taking many industrial secrets with him. He was instrumental in designing and building the first successful cotton-spinning mill in the US in 1792 at Pawtucket, Rhode Island. Later, President Jackson referred to him as "the father of American manufacturers." Unsurprisingly, he is known over here in the UK as "Traitor Slater."

However, it is one thing to use a warm air system in a factory where labor can be employed to operate it. It is quite another to adapt it for a house. This only became possible with the invention of the electric fan by Schulyer Wheeler (originally of New York State) in 1886. In a modern domestic installation, a fan blows thermostatically controlled air from the furnace through a filter into ducts that lead to each room. The air enters the rooms through adjustable registers, and the cooler air in the rooms is pushed back to the heating unit through return ducts. The air therefore carries heat energy constantly and cyclically around the building.

Strutt's system relied upon a stove, which would be no more than 30% efficient, if that. Modern furnaces do very much better, although any installed prior to 1992 should be replaced, since from that date a minimum AFUE (annual fuel utilization efficiency) of 78% for all new fossil fuel furnaces has been required. It would not be unusual to find an efficiency of only 60% for earlier furnaces. The recommended (though not required) level is now 90%, with 97% the best achievable (see www.eere.energy.gov/femp/pdfs/gas_furnace .pdf). The sizing of the furnace is also important—an oversized furnace can be very wasteful.

The 90% or more AFUE ratings are achieved by *condensing furnaces*. These recover much of the 20%–30% of heat energy normally lost through the flue. You may remember that just as evaporation absorbs heat energy to produce cooling (the science behind sweating), condensation releases it. In a condensing furnace, condensation is deliberately induced in the water vapor produced by combustion to release the energy that would otherwise be lost. An additional heat exchanger is incorporated into the furnace, which therefore tends to be bulkier and more expensive than a conventional one. The time taken to recover the extra outlay depends on the local climate, so you need to get estimates or do your own before proceeding. Irrespective of the cost, you will reduce CO_2 emissions.

Yet there are problems with condensing furnaces. You cannot, for example, simply replace an existing conventional furnace with a condensing one. It's not like changing your automobile. It's more like changing from an automobile to a personal helicopter! Provision has to be made for the condensate to be collected and drain away without any danger of freezing. Also, the flue gases are cooler and therefore less buoyant. Additional condensation could therefore occur in the flue. This will be acidic and likely to attack the inside of the flue, which therefore must have an appropriate flue liner. Once again, specialist advice is needed.

If you think of the furnace and fan as the heart of the system, then the ducts are its arteries and veins—and are just as important. Don't let their apparent simplicity lead you to take them for granted. As discussed in Chapter 3, they must be properly sealed and insulated to reduce heat energy losses (or gains for cooling systems). An incorrect layout can also have an adverse effect, especially if it leads to lots of twists and turns in the ducting that impede the airflow. The airflow can also be restricted if the return ducts are not big enough, because of bad design, for example.

Balancing the settings of the registers can also be a problem. Figure 6.2 shows a centrally mounted furnace unit typical of a small apartment. This has short stub ducts opening into three rooms. The thermostat is fixed to the wall in the living room (on the right in the diagram). Surprisingly, partially closing the grid in this room does not alter its temperature. What happens is that the thermostat keeps the fan and furnace working harder to maintain the same temperature. This means the other rooms receive more hot air through

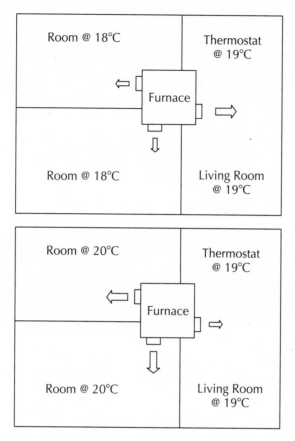

FIGURE 6.2.   A small forced air heating system. *Top:* open register in living room. *Bottom:* partially closed register in living room. Arrow size indicates comparative flow rates.

their ducts and their temperature rises, wasting heat. The same can happen in larger systems, though perhaps not so obviously.

Working out the best pathways, settings, and sizing is a specialist job, particularly as every home is unique. In general, therefore, the homeowner should confine herself to changing the filter and cleaning the fan blades and registers every month during the heating season. An annual maintenance visit from a qualified engineer is recommended practice. Additionally, if anyone in the house has allergies or asthma, a better filter may be needed. Jeffrey C. May, in his book, *My House Is Killing Me!* points out: "Many people think the purpose of the furnace filter is to clean the house air. The typical fiberglass air filter found in most hot air systems was designed to keep paper and other debris away from the heat exchanger, not to keep the house air clean" (p. 177). For better air quality, he recommends a media filter. The extra comfort more than repays the extra cost.

The small, unobtrusive registers of warm air systems have an additional unexpected benefit. I have a fond childhood memory of an uncle and aunt buying a new bookcase. Unfortunately, when it was delivered, the only place it would fit was in front of a radiator (they had a hydronic system). For an hour or so, it was like Laurel and Hardy—furniture being moved around and getting stuck, people banging their fingers, both blaming each other. I went home to tell Mom and Dad, who fell about laughing. Registers, being much smaller than radiators, allow greater freedom of furniture positioning.

Another problem for my aunt was that some of her furniture developed cracks due to the frequent humidity changes characteristic of the UK climate. A warm air system might have solved this too, especially if it had a humidifier feeding humidity-controlled air into the ductwork. Sadly though, fitting a new warm air system would not have been an option, since floors would have to be lifted and relaid to accept the ducts. Warm air systems are generally recommended only for newly built houses, rather than as a retrofit.

There are other problems too. For example, there is no radiant heat emitted, which means that the output air must be a degree or two higher than it would otherwise need to be (see Chapter 1). The ducts also provide a warm, dry habitat for pests such as insects or small mammals. The very thought of spiders crawling out at night would have been enough to give my aunt nightmares. Likewise, noise and odors can move from room to room through the ducting—music played in bedrooms may be heard throughout the house, and the smell of onions frying in the kitchen can appear in the bedrooms. Best to say nothing about bathrooms!

A useful fact sheet on warm air heating is available from BC Hydro, downloadable from www.bchydro.com/rx_files/pshome/pshome1597.pdf.

Pumping the Heat

These days, my favored form of cardiovascular exercise is walking in the countryside near my home—quite a difference from forty years ago when I represented my university at cross-country running. I recall that our coach was adept at devising various circuit-training "tortures" for our evenings in the gym, many designed to get our hearts pumping our hot blood more efficiently. The same principle of moving heat energy from one place to another using a circulating fluid is utilized in a heat pump, which has the added advantage that you get more energy out than you put in. I could certainly have used some of that when I was running competitively!

The secret of this apparent "free lunch" is that energy is being moved from one place to another rather than coming directly from a fuel of some kind, as is the case in a furnace or boiler. It may not seem so in the depths of winter, but there is always some heat energy available in the air. An air-source heat pump takes some of this to warm up buildings, leaving the atmosphere slightly colder. If this seems odd, think about your refrigerator. Have you noticed that when you come back from the store and load it up, there is a flow of warm air from the back for a while? What has happened is that heat energy has been taken from the food, and thrown out at the back of the machine. Colloquially, heat has been "pumped" from one place to another. With a heat pump, for an expenditure of 1 BTU of electrical energy, up to 4 BTUs of heat energy can be delivered from the outside atmosphere into your home—an energy efficiency of 400%. It is more usual, however, to refer to a heat pump's coefficient of performance (COP) rather than efficiency, though they are effectively the same thing. In the example just given, the COP is 4.

Our first refrigerator lasted us nearly twenty years, much to the despair of my wife, who prefers to have the latest design and styling. By contrast, my philosophy is "If it ain't broke, why change it?" and we agreed to wait for it to break. But why did it last so long? The simple answer is that there is, in fact, very little to go wrong in a refrigerator because of the relatively few moving parts involved.

Figure 6.3 shows the principle. A liquid is pumped cyclically around a circuit of pipes, part of which is narrowed. This part of the tubing is called

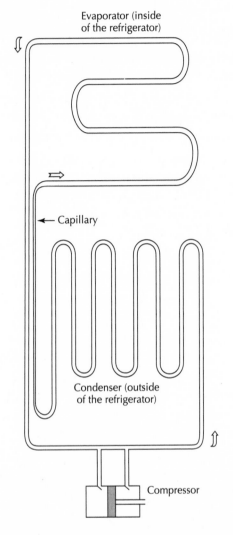

FIGURE 6.3.   The working parts of a refrigerator

a capillary. When it leaves the capillary, the liquid is forced to evaporate to become a gas—a bit like people spreading out in all directions as they leave a ball game through the relatively narrow exits. This produces cooling in the gas, just as evaporating sweat reduces temperature (see Chapter 1). This bit of pipework is inside the fridge, so heat energy from the warm food flows into the cool gas.

The cool gas is forced to condense back to liquid in the pipework *outside* the refrigerator (as it enters the capillary) releasing heat energy into the atmosphere, and "presto," energy has effectively been pumped from inside to outside the refrigerator! An internal thermostat is fitted, which switches the pump on and off to control the temperature of the food, and the refrigerator is heavily insulated to prevent heat entering it from the surrounding air. Apart from corrosion, only the pump and thermostat can go wrong—both of which are reliable and easily replaced. Twenty years is not unusual for the lifetime of a refrigerator.

This reliability, however, has come at a price: in the early days refrigerators could leak toxic gases and poison the owners. This happened to a German family in 1925, an event that started none other than Albert Einstein thinking. He teamed up with Leo Szilard, and together they designed a leak-proof refrigerator without moving parts. Unfortunately the project ground to a halt in 1930, when other American inventors came up with a new non-toxic refrigerator gas called *freon*. The full story is told in Steve Silvermans's *Einstein's Refrigerator and Other Stories from the Flip Side of History* (2001). Freon is now known to destroy the earth's ozone layer, and special arrangements are therefore needed for the disposal of old fridges. Consequently, there has been a small revival of interest in Einstein-Szilard refrigeration.

Heat pumps are often referred to as "refrigerators working backward." The energy required is taken from convenient surroundings—not necessarily the air, for a nearby river or the ground will work as well—and moved to the inside of the building. Heat pumps can therefore replace the furnace of forced hot air systems. One of the first buildings to be heated in this way was the Royal Festival Hall on the banks of the river Thames in London, built for the Festival of Britain in 1951. It took its heat energy from the river and sent the water on its way only very slightly colder, though the press at the time ran stories about the possibility of the Thames freezing over. The pump is still working, and it ain't happened yet.

Is this feasible for an ordinary house? Systems are available for homes with a suitable nearby river, though taking the energy from the air (an air-source heat pump) or the ground (a ground-source heat pump) is more common. One problem with air-source heat pumps is that, as with refrigerators, ice can form on the cold pipework and they may require defrosting from time to time. However, a small electrical heater can be incorporated to do this automatically.

The COPs of heat pumps vary with temperature. Without going into too much detail, this is a consequence of the second law of thermodynamics. Provided the temperature in the house is constant at around comfort temperature (65°F), the COP is inversely proportional to the temperature difference between inside and outside. So, when the temperature outside drops and the temperature difference goes up, the COP falls. This is not surprising, since we would intuitively expect it to be harder to extract heat from something that is already cold. It's rather like me taking a hot drink in my flask when I go walking in the hills. It warms me up because my body extracts the heat from it much more effectively than from a lukewarm drink—there is less heat there to extract.

Typically, COPs drop considerably once outside temperatures go below 15°F. Consequently, heat pumps are not really efficient in northern states, and even elsewhere a supplementary backup system may be needed for very cold days. People buying an air-source heat pump should therefore look closely not only at the heat pump's COP but also the outside temperature range for which the quoted COP is effective.

A ground-source heat pump overcomes this problem because it draws heat energy from the ground, which below a depth of about eight feet, is at a relatively constant temperature all year round. Its COP is consequently more stable and higher than that of an air-source heat pump for any given location. You rarely, however, get something for nothing. The price for this improved performance is that it is significantly more expensive to install.

This is all very well, but if you live in one of the southern states, pumping heat energy *into* a building must seem a very strange thing to do—a central cooling system would make more sense. The good news is that heat pumps can just as easily produce cool air as warm air, using the same system of ducts and registers. They can even cope with a climate that demands hot air in the winter and cool air in the summer, making them extremely versatile. They are also at the heart of air-conditioning units. Further information on heat pumps is available at www.bchydro.com/powersmart/elibrary/elibrary683.html and

www.eere.energy.gov/consumer/your_home/space_heating_cooling/index
.cfm/mytopic=12610.

Air-Conditioning

In the eighteenth century, Newgate Prison in London had a reputation second
to none for death and misery within its walls. The deaths often had nothing
to do with public executions—though there were plenty of those—but were
due to "gaol fever," which we now know to be typhus. Drawings of the prison
around 1800 show a vertical axis windmill on its roof. What for? It was part
of the prison's ventilation system, installed not only because prisoners were
dying of the fever but also because jailers and judges met the same fate. The
windmill blew clean fresh air down a shaft, which ventilated the cells and had
a cooling effect, though then, as now, health was considered more important
than comfort.

Technology has moved forward considerably since then. As far back as
1902, the "new" New York Stock Exchange building incorporated both a cooling and a heating system for the trading floor, with the cooling system based on
refrigeration technology. Today's heat pumps operate the refrigeration cycle
of figure 6.3 and, as part of an air-conditioning unit, can service a single room
or a whole building. The condenser coils face the outside as they do for a refrigerator, and the extracted heat energy is scattered into the atmosphere. The
cool evaporator coils are on the inside of the room, with a separate fan to push
the cooled air into the room or into the ducting.

AC units need, just like refrigerators, to be defrosted from time to time.
This was one of my chores with our first refrigerator and the "price" I paid for
refusing to buy a new one until the old one wore out. I often forgot and wound
up carefully chipping ice off the evaporator coils. Mysterious pools of water
also seemed to appear on the kitchen floor as soon as my back was turned.
Needless to say, our second refrigerator had automatic defrost. With AC units,
moisture from the room condenses on the evaporator coils. It usually drips
harmlessly into a small tank at the base of the unit, which must be emptied
from time to time. Alternatively, a drainage pipe leading to outside drains can
be fitted to the unit. In some conditions though, it can freeze, and so a defrost
mechanism is needed for AC units too.

Some modern units are fitted with a reversing valve that can change the direction of the refrigerant flow. Reversing the flow for a short time defrosts the coils (other designs have a small electric coil). The reversing valves allow these units to provide heating in summer as well as cooling in winter in mid-latitude states.

There is more to air-conditioning than blowing cool or warm air. The air should also be cleaned. The easiest way to achieve this is by filtering the air prior to its entering the room. In early systems, cheesecloth was used, but today much more sophisticated and effective filters are available (see again Jeffrey C. May's *My House Is Killing Me!*, p. 177). Yet temperature control plus filtration is still not enough to ensure indoor comfort. You may remember from Chapter 1 that the right humidity is also important for comfort. Gail Cooper, in her excellent and readable book *Air-Conditioning America* (1998), remarks: "The most frequent complaints about indoor comfort are dry air in the winter and humid air in the summer. Cold air, with its small capacity for moisture, has a low absolute humidity; thus, when it's brought indoors and heated, the relative humidity plummets as the temperature rises. The resulting hot air is dry, and it dehydrates both people and furnishings" (p. 11).

Humidifying the air is not difficult; a suitably placed saucer of water will help. I have also seen various devices that rely on a wick in a reservoir of water—sometimes with a fan. When the NYSE building was designed, it had a humidifying system consisting of steam pipes immersed in a water bath to produce the necessary evaporation. A more modern system consists of a pad mounted in a rotating cylinder, partially immersed in a water bath. A fan then passes air over the pad, which evaporates the water and carries it into the room. This has the advantage that switching on or off the electricity supply to the fan can control the humidity.

Dehumidifying air is more difficult. Gail Cooper writes of Alfred R. Wolff, the designer of the NYSE system: "He calculated that he would need 119 tons of cooling capacity for cooling the air and 180 tons for extracting the moisture from the trading room . . . it can perhaps be called the first air-conditioning installation because of its self-conscious attempts to control all four key factors: temperature, humidity, cleanliness, and distribution of air" (p. 15). Thus, removing moisture requires much more energy than cooling. In most parts of the US, the moisture removed by the air-conditioning unit is generally sufficient to ensure comfort, since the human body can tolerate a wide range of humidity levels (see Chapter 1). This is not sufficient, however, in the warm, humid climates of the southern coastal states. One way of coping with this is to repeatedly return air through the AC unit—effectively wringing it out, just like my mother used to do with the weekly washing. The cost of doing this escalates alarmingly.

Full control of humidity is achieved by a fitted humidistat. Just as a thermostat switches a power source on or off, a humidistat can turn on or off humidifiers and dehumidifiers in response to humidity. So, the temperature, humidity, cleanliness, and distribution of air are all controllable. In one of his lesser-known plays, the Bard of Avon has Guiderius sing:

> *Fear no more the heat o' th' sun*
> *Nor the furious winter's rages:*
> *Thou thy worldly task hast done,*
> *Home art gone, and ta'en thy wages.*
> *Golden lads and girls all must,*
> *As chimney sweeps, come to dust.*
> —William Shakespeare, *Cymbeline*, IV:2

Thanks to air-conditioning we no longer have to await death to be freed of the fear of the sun's heat or the fury of a winter storm. Mind you, the future is not all rosy for the air-conditioning industry. Studies in California have shown that ventilation cooling *alone* in a well-designed house can achieve a high level of comfort, which threatens to make air-conditioning obsolete in some areas (see, for example, www.homeenergy.org, particularly the July/August 2003 issue).

However, ventilation cooling is not yet a possibility for most homeowners—few of us can afford a house designed to our own specifications. As always, energy efficiency of our AC equipment is a major concern. For homes, this is most commonly measured by the seasonal energy efficiency ratio (SEER), the ratio of the cooling capacity (in BTU per hour) to the power input in watts. Current standards are a minimum of 13 for appliances manufactured after

January 23, 2006. Many older systems only reach 6, so you could potentially halve your cooling costs by updating you existing unit.

Getting the right size and model for your particular home and climate is as important as the SEER. The optimum positioning within your premises is also important. It is always worth getting professional help and advice, once you have a reasonable background understanding of what you want. Informative and well-written fact sheets for both heat pumps and air-conditioning are available from the US Department of Energy at www.eere.energy.gov/consumer/your_home/space_heating_cooling/index.cfm/mytopic=12370.

Steam!

What is it about steam? Is it the sense of raw power? The sense of history? The smell of hot steam and lubricating oil? I don't mean going for an afternoon ride on a steam locomotive—or watching a traction engine plow a field. Weekend after weekend there are people who go to their local railroad museum to engage in the sheer hard work of restoring locomotives, carriages, and track as unpaid volunteers. Some even pay out good money simply for an afternoon on the footplate of a steam locomotive. In the UK, steam is becoming big business. We have 108 operating railways and 60 steam centers, employing over 1,000 staff and utilizing 2,300 volunteers (see http://ukhrail.uel.ac.uk/facts.html). Americans are perhaps a little less eccentric, but the phenomenon certainly occurs in the US too—see, for example, www.steamlocomotive.com/.

There's an old saying that "children keep you young." When my granddaughter reached 7 years old, I had the great joy of taking her on her first-ever ride on a steam train. I spotted the radiator shown in figure 6.4 in the corridor of our carriage and duly photographed it. The conductor informed me that it ran off steam from the engine. I was skeptical. We were about five carriages back from the engine, and the radiator must have been below the level of the engine's boiler, so I couldn't figure out why it didn't fill with water. Unfortunately an engineer wasn't available to answer my more detailed questions—and I never did find out for sure how it worked.

To enthusiasts, having a steam boiler in the basement "driving" a heating system must be a welcome reminder of their hobby. But alas, steam-operated heating systems are yesterday's technology—just like steam-operated railroads.

Figure 6.4.   Radiator in train carriage (c. 1950)

They are rarely installed now in residential buildings, and existing ones are often converted to hot water (hydronic) distribution. They do, however, warrant a brief mention here.

A steam heating system consists of a boiler and insulated pipework to carry the steam to the radiators, where it condenses and gives up its heat energy. It then returns to the boiler as condensate before being reheated and sent around the cycle again. The distribution layout varies in complexity and in maintenance requirements from system to system. The radiators must always be positioned above the boiler (otherwise they fill with some of its water) and be accurately pitched to drain toward it. At first use, the system will be full of air, and as the system continues in use, air may enter the system from that dissolved in water. An air valve is therefore incorporated in the radiator to allow it to be vented to the atmosphere. This valve should be cleaned from time to time.

Even from this highly simplified account, it is easy to see that there is much that can go wrong and that there is a need for skilled preventative maintenance. In addition, steam-heated buildings often have the problem of uneven heating, together with annoying intermittent thumping noises (pounding). These can usually be minimized by making small adjustments to the system, but this is really a specialist job. Therefore, if you are a steam enthusiast, you can do little to indulge your hobby at home, other than checking the water level in the boiler from time to time—unless, of course, you lay a miniature track around your garden and build (or buy) your own small locomotive!

Since hydronic systems tend to be simpler and more energy efficient, it is not difficult to see why they often replace steam systems, which continue to decline in popularity. The joy of tinkering with steam engines rarely extends to tending an inefficient heating system at home.

Pumping the Water: Hydronic Systems

Bill the Boiler Man (Gravity Systems)

My earliest ambition was to have a job like Bill's; as a boiler man, he kept our school's heating system working. I looked forward to school because, however bad the weather, the central heating kept us warm, for he fed the boiler with huge shovelfuls of coal as needed. His boiler room was also used for storage, and from time to time pupils were sent to fetch things. It was a warm, dry place, which Bill had gone to some trouble to personalize with a small table, kettle, cupboard, and old but comfortable easy chair. If he left the door of his cupboard open, you could see pictures of scantily clad young ladies, which reduced many a lad into fits of giggles.

Bill's working hours were attractive, too; he went home at about 1 pm and in the summer worked outside on the school grounds. What I didn't realize was that he had to start at about 4 am to get the boiler going. Nonetheless there were long periods of the day when he could sit, read his paper, and let the boiler boil. What bliss!

The kind of system over which Bill had charge was an early hydronic system. Hot water, produced by a central boiler, passed through huge pipes to each room. The hot pipes went all around the walls of our classrooms and warmed the whole room. It was then returned to the boiler for reheating. The circulation of Bill's boiler was achieved by natural convection through large-diameter pipes. This is known as a *gravity system*.

Convection is a very slow process, hence the need for Bill to start so early in the morning. Heat energy escaped from buildings throughout this long warmup time, making such systems inefficient. Gravity systems are no longer installed but can still be found in some older buildings. They are nevertheless a good starting point for understanding modern hydronic systems.

Better and Better (Pumped Systems)

A major improvement to Bill's system incorporates an electrically operated pump to move the hot water around more quickly through narrower pipes, providing shorter warmup times and greater economy. Better control is also possible, since the pump can be switched on or off automatically by a thermostat or a timer. Alas, this has made boilermen largely redundant, and I couldn't have fulfilled my childhood ambition even if it had remained unchanged.

Boilers have changed out of all recognition since those days. They are still a heater and water container, but massive improvements have been made to their control. Even small domestic boilers have automatic ignition and thermostatic control, making manual intervention largely unnecessary. High-efficiency condensing boilers are also available that perform in a similar way to condensing furnaces, with the same potential problems. Both types of boilers need annual cleaning and maintenance inspections by a qualified person. Like tuning your automobile, it's a specialist job.

From the boiler, the hot water is pumped around a circuit with radiators at appropriate points. The American William Baldwin invented the first practical domestic radiators in the late nineteenth century—necessary because narrow pipes have insufficient surface area to be good heat emitters. These radiators were similar to the one shown in figure 6.5, which is in Bramall Hall, an Elizabethan "black and white" house in Stockport, England (obviously, the radiator was a late addition). Pressed steel panels welded together (figure 6.6) have now largely superseded the original cast-iron design, though the latter is still available for that retro look. The downside of radiators is that internal corrosion eventually leads to leakage. This will show itself initially as seepage, rather than a massive burst, and typical radiator lives are well in excess of ten years (mine are still showing no sign of leaking after thirty years).

Radiators are misnamed, for most of their energy is given out by convec-

FIGURE 6.5.   A Victorian cast-iron decorated radiator

tion, the proportion depending on the radiator design. Modern designs have unobtrusive fins—like those on a motorbike engine or at the back of a refrigerator—welded to them to greatly increase the area for releasing heat energy. Warm air rises from the radiator, cools, and drops to set up a cyclical movement back to the radiator. Thus, space is needed for an easy airflow around them. They therefore need to be mounted slightly away from the wall and above the floor, which makes them difficult to clean and decorate.

The placement of radiators relative to windows and doors is also important. They are generally fixed below windows to allow the rising warm air to counteract any cold air dropping from the windows. Drapes covering radiators are not a good idea, for they will restrict the airflow and direct it towards the cold window. A radiator shelf on which shorter drapes rest is a good solution, for they stop warm air from being trapped between them and the window, as well as saving money on the amount of material needed.

Radiators are fitted with valves at each end so they can be easily removed from the wall when decorating or if they start to leak. One of these can be an *automatic zone valve* (also known as a *thermostatic radiator valve*), which shuts off the water when the room temperature is too high and turns it on when it is too cold. One can be seen at the bottom of the radiator in figure 6.6. This is much easier than fiddling with the registers of a warm air system to give each room its own temperature. It is also much more sensitive than having one thermostat for the whole house operating the pump. Not much use, though, if you leave all interior doors open!

I once dipped into an old medical textbook and was surprised to see how many ailments were treated by bleeding the patient, a practice that often did more harm than good. It made me feel ill just to think about it. It may surprise you to know that radiators need "bleeding" too. Air dissolved in the water forms tiny bubbles of gaseous air when the temperature rises. Eventually this collects at the high points in the system reducing its efficiency. There is usually a *bleed valve* at the top of a radiator (also visible in figure 6.6). Releasing this with a radiator "key" lets the air out, but be careful if you try it yourself, for it is apt to spit mucky water out. There may also be other bleed points on a system depending upon its design.

One way in which the efficiency of a radiator can be improved is by making better use of the radiant heat it gives out; remember, this form of heat travels in straight lines like light and heats up things in its path. About half of a radiator's energy is given out in the direction of the wall behind it and passes through to the outside, wasted. Cheap radiator reflectors can be fitted behind the radiator to reflect the heat back into the radiator, saving both money and energy. Simple but effective.

Radiators do take up a lot of wall space, restricting the arrangement of furniture—remember the Laurel and Hardy situation my aunt got into because of this? This can be overcome by using other heat emitters such as baseboard convectors instead. These are simply long lengths of pipe through which the hot water flows, with fins welded directly onto them. They have sheet metal covers to make them less obtrusive, and louvers to allow the rising warm air to escape into the room. They are placed low along a wall and can be recessed into the floor. Just as for radiators, a good airflow is necessary, so the louvers must be kept clear and clean. Vacuuming them at the same time as the carpet will usually suffice.

FIGURE 6.6.   A modern pressed steel radiator

Cabinet convectors are another option. These large, louvered cabinets contain one or more finned tubes. As with baseboard heaters, the louvers must be cleaned and unblocked. They should be carefully positioned in the same way as radiators. Personally, I don't see much advantage over ordinary radiators, but if you prefer the design then go ahead. They work just fine.

When I was very young, I remember being given a bag of small chocolate Easter eggs at a friend's house. We decided to play with them a bit before eating them and had fun lining them up on top of a radiator. You can guess what happened next: we went to watch the TV, the heat came on, and I was left with a gooey mess. Radiators, convectors, and baseboard heaters all have an operating temperature of around 180°F (dangerous if touched), because convection needs a large temperature difference to drive it. You don't get that kind of problem with warm air systems.

Before I was born, my grandmother had a kitchen range, which supplied all the household's domestic hot water from one outlet. Modern hydronic systems can supply domestic hot water for a number of outlets as well as supplying hot water for the radiators. This is achieved by incorporating an insulated hot water storage cylinder into the circuit. This cylinder contains a heat exchanger consisting of a copper coil with connections to the outside. Figure 6.7 shows the idea. Hot water from the boiler passes through the coils and then is returned to the boiler for reheating. This process indirectly heats the water surrounding the copper coil, which is stored until required. This type of hot water cylinder is sometimes called an indirect tank. There is another circuit around the radiators attached independently to the boiler, so domestic hot water and heating water can be controlled separately. Because the water expands as it is heated, another small tank is needed to receive the expanded water.

I believe that most people these days are so busy and their world so complex that they yearn to keep things simple—I certainly do. Why not then simply place a heat exchanger coil inside the boiler? You can: such an arrangement is known as a tankless coil. The bad news is that tankless coils cannot cope with today's demands. They worked in homes where there was one tub and a couple of hot water faucets, but would be underpowered for today's homes that also have washing machines, dishwashers, shower units, and more.

There are other ways of supplying domestic hot water. I am particularly impressed by the simplicity of on-demand electric water heaters, which are fitted with high-power heating elements. They are the big brothers of the humble kitchen kettle. A full discussion of domestic hot water provision is beyond the scope of this book, but there is plenty of useful, up-to-date information on the Internet if you wish to pursue the topic further. See, for example, www.eere .energy.gov/consumer/your_home/water_heating/index.cfm/mytopic=13 020.

In spite of their various advantages, there's no getting away from the fact that hydronic systems are more complex than warm air ones. They can, however, be fitted pretty much anywhere. Various levels of control can be built

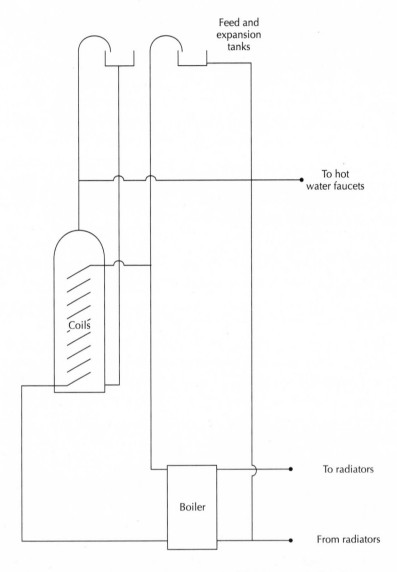

FIGURE 6.7.   Typical layout of a hot water and heating system. Overflow pipes
(not shown) take expanded water to the outside of the building.

into them, and the more sophisticated ones allow the hot water and heat to be controlled separately with respect to both time and temperature.

Bill the boiler man would be amazed at what can now be done!

Up Above and Down Below (Hydronic Radiant Systems)

You may remember Mel Gibson's 1995 film *Braveheart*, based upon the rebellion of the Scots led by William Wallace against the English. There was considerable ongoing violence between the two nations throughout the thirteenth and fourteenth centuries, subsiding only when the two kingdoms united in 1603. Even after this unification, cross-border raids still occurred. If you lived in the border region of Northumbria, your main priority would be to defend your home.

Bastle houses were defended, stone-built farmhouses of two stories, common in Northumbria. The ground floor housed animals, which were brought in for the night while the family slept upstairs. Should raiders appear, they would have to push their way through cows and pigs, risking injury and awakening the household. One can only imagine the stench that these animals would generate crammed into such a small area. They would also generate much body heat, which would permeate upward, warming the living quarters—an early example of underfloor heating. (More detail can be found on the website www.reivers-guide.co.uk/borderreivers3.htm.)

Laying stiff plastic pipes within the concrete slab of the floor as the house is being built and pumping warm water through them can achieve underfloor heating much more conveniently. The whole floor slowly releases heat energy, which rises upward without restricting furniture position in any way—no Laurel and Hardy scenarios! The floor surface temperature only has to be 7°F above the room temperature for convection to start, so it won't burn your feet. Floor coverings are too thin to block the heat from being emitted. Moreover, a significant component of the heat is radiant, giving a more natural feel to the heat. (Remember Chapter 1, "Feeling Good"?) Zone control can be achieved simply by having separately controlled loops of piping for each room. As warm air rises, upper rooms are also heated, though heat can take an hour or more to reach them. Radiant floor systems are not suitable for suspended wooden floors.

Radiant ceiling systems are another option and can be retrofitted relatively easily. Tubing is laid just above the gypsum sheet boarding of the ceiling with insulation above that, to reduce the upward flow of heat energy. The overall loss of space is about 1.5 inches. Hot water, typically between 100°F and 120°F, is circulated through the tubing, and the entire area becomes a radiator. The radiant heat is directed downward, so the head is warmed more than the feet. Sitting at a table can be a problem because, just like sitting under a tree on a sunny day, radiant heat from above hardly penetrates through to the lower half of the body. This shadow effect can mean that you may finish up with cold legs and a hot head!

More details can be found at www.eere.energy.gov/consumer/your_home/ space_heating_cooling/index.cfm/mytopic=12590.

Waste Not, Want Not

This proverb has a particular poignancy for me and many other Brits of my generation. Born in the middle of World War II, we quickly learned to waste absolutely nothing. We bathed in six inches of water, saved all our paper, mended all our clothes, and generally did without. Failure to eat all the food put before us was a crime punishable by being sent to bed without any supper for three nights in a row. Old habits die hard, and now my wife despairs of me, for I still find it hard to throw things away before they are properly worn out (for a shirt, this means at least two holes of a minimum half-inch diameter).

All of the heating and cooling systems described so far are inherently wasteful. They rely on fuels being delivered to individual households where they are either converted into heat energy or used to produce cool air. This means there has to be a boiler, furnace, heat pump, or other appropriate plant on the premises. Compare this with district schemes, which distribute hot or cold water from a central point through a network of pipes to individual homes. Such schemes not only reduce the capital, operating, and ongoing costs for householders but also leave them with more useable internal space in their homes. Furthermore, because of the high efficiency of the central plant, fewer greenhouse gases are emitted overall.

In some district heating schemes, coal or oil is simply burned in a large boiler. However, if the geology of the area is right, free hot water from underground can be utilized. The state of Idaho has been doing this since 1892, when the Boise water works was opened (see www.eere.energy.gov/geo thermal/pdfs/directuse.pdf and www.idwr.state.id.us/energy/alternative_fuels /geothermal/district.htm).

For cooling systems, cold water can be extracted from the bottom of a lake, suitable if the water resides there at a constant temperature of about 39°F. It is then purified and fed through a heat exchanger, where it cools water in an adjacent network of pipes that supply individual buildings. Toronto is operating such a system using Lake Ontario, and the company behind the scheme claims that only a quarter of the energy needed for conventional air-conditioners is used to run it. So not only is there a cost savings but less carbon dioxide is released into the atmosphere, and because significant demand is removed from the already overloaded grid system, outages are less likely.

Another energy-efficient source of heat energy is the local power station. Thermal stations burning fossil fuels operate at efficiencies of 40% or so at best, since a lot of the heat energy goes up the chimney of the boiler and out through the cooling towers. Power stations can be modified to cogenerate electricity *and* hot water for distribution around a district. Such schemes are called combined heat and power systems, or CHP for short, and can reach efficiencies of 60%–90% overall. The Massachusetts Institute of Technology has a 22 megawatt CHP station, which supplies around 95% of its power, heating, and cooling needs. Their annual electricity bill for 2000 was estimated to have been reduced by around $5.4 million by this system. Is it any wonder that it is now federal policy to expand this provision by making information and technical assistance available to federal agencies? See www.eere .energy.gov/femp/technologies/derchp_chpbasics.cfm.

With such "big bucks" savings achievable, you might well ask, "If it's that simple why isn't everyone doing it?" Well, there is a major problem providing the infrastructure—imagine the cost of laying insulated underground pipes to each and every house or building in an area. It would only be worthwhile if a very large proportion of owners accepted it. But there is a Catch-22 situation here. Owners cannot make a decision until they have seen the ongoing prices, and the companies cannot price the system until they know how many will use it. CHPs have therefore been most used by academic institutions, hospitals, correctional facilities, and military installations, where users have little choice in the matter.

Some areas are prone to the loss of electrical power. Fortunately incidents tend to be of short duration, but nonetheless they are a real nuisance when they happen. I have learned to find my way around our house blindfolded, a consequence of having fumbled around in the dark looking for candles on occasions when the wind has brought our power lines down.

The ideal solution would be to own my own power station, but until now my budget couldn't accommodate this. Today a new micro-CHP unit is being test marketed in the UK that will generate an electrical output of 1.2 kW with a thermal output of 8 kW. It is based on an almost silent gas-fired Stirling engine about the size of a small washing machine and is manufactured by Whisper Gen of Christchurch, New Zealand. My guess is that it won't be long before they become available in the US. Links to this and similar products can be found at www.energy.ca.gov/distgen/equipment/stirling_engines/vendors.html. Certainly worth considering next time you need to upgrade your system, but get some independent test reports first!

Faraday's Legacy

As a young physics student, I regarded Michael Faraday with a mixture of awe and dread. The major achievement of this son of a London blacksmith, born in 1781, was to lay the foundations for all future electrotechnology. My dread arose out of the need to understand his theories and laws. He discovered electromagnetic induction (and therefore generators), a fair chunk of electrochemistry, and much else. Electricity is his bequest to us all. Mind you, its value was not apparent at the time. His contemporary, Prime Minister Gladstone, queried him about the usefulness of electricity. Faraday's famous (and perceptive) response was, "Why sir, there is every possibility that you will soon be able to tax it!"

If I travel to the US, I have to change my UK pounds into US dollars, and then when I come home I have to change them back again. I can easily lose 5% of the value of my money at each stage. Using electricity for home heating is a bit like that. When it comes from thermal power stations (as opposed to wind turbines or hydro schemes, for example), heat energy is changed to

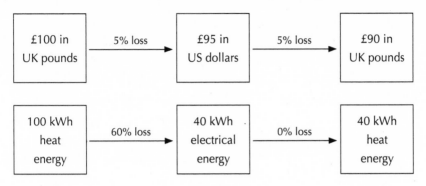

FIGURE 6.8.   A comparison of "exchange rates" for currencies and energy

electrical energy at an exchange rate (at best) of 40%. When changing from electrical to heat energy, there are no "exchange rate" costs, but overall there has been a 60% energy loss (see figure 6.8). So don't be fooled when you are told electricity is 100% efficient; this is only true for the final stage, changing the electricity back to heat. Electrical heating tends to be expensive in terms of both energy costs and environmental impacts. Nonetheless, its ease of control using switches, thermostats, and timers makes electrical systems popular. As with electric fires, there are a number of different options.

Electrical Baseboard Heaters

Electrical baseboard heaters are amazingly simple, consisting of heating elements mounted inside a long thin metal box. As with hydronic baseboard heaters, louvers near the top and bottom allow air to flow over the elements to emerge at the top at a warm comfortable temperature. Thermostatic control prevents overheating and allows for economic operation. They are effectively small versions of the electric panel heater shown in figure 5.7. As such, they are unobtrusive, take up little space, and come in a variety of shapes and sizes. The only maintenance needed is to vacuum the grills regularly to keep them clear of flammable dust and fluff. More detailed information about electrical baseboard heaters is available at www.bchydro.com/rx_files/pshome/pshome1595.pdf.

Up Above and Down Below Revisited

Ceiling and underfloor heating have already been discussed in relation to hydronic systems. In electrical systems, the hot water pipes are replaced by heating elements. The two systems have much the same strengths and weaknesses.

There are two important differences, however. First, it is possible to buy both floor and ceiling tiles that have electrical elements embedded in them. They can therefore be easily fitted into an existing building. The second difference concerns underfloor heating. Some electricity companies structure their rates to charge less at night. These time-of-use (TOU) rates even out the demand, so that power stations don't have to keep stopping and starting, thereby making it economic to offer these rates. Underfloor heating is ideal for taking advantage of the rates, since it can "charge" up the concrete slab with heat energy during the night at the cheap rate and then slowly release it during the day.

Timers cannot guess how long is needed to "charge" up the floor to store enough heat energy for the next day, since they cannot predict tomorrow's temperatures. Provision is therefore made to allow daytime topping up if insufficient heat has been stored during the night. Either a manual override can be used to switch the heating elements on for a while, or a thermostat can do this automatically. There is, however, a potential wastage problem with storing too much. Toward the end of the war, Mom got adept at beating the shortages. She bought stacks of stuff whenever it was available and stored it wherever there was space. We ended up with far more than we could use, and indeed much had to be thrown away. In the same way, people with any sort of storage heating tend to set their timers to store more than they need to be sure there is enough, with consequent inevitable waste.

Lightweight ceilings with embedded heating elements are unable to store heat, so the electric current has to be used at the full rate, which makes them expensive to run. They can be fitted into an existing home relatively easily and are particularly useful for upper rooms where the ground floor is the main source of heating and a quick response is needed.

A Switch in Time Saves a Dime

But what if you do not have a concrete slab floor? A switch to a time-of-use tariff can still lead to significant savings, but only if you invest in the right equipment, such as electrical thermal storage (ETS) radiators. My mother-in-law had ETS radiators installed in the early seventies, and I remember them as large, thick, heavy, and wall mounted. In fact, when I first saw them, I wondered if the wall would carry their weight. Today they are much thinner and lighter. Interestingly, the evolution from thick and heavy to thin and light was made possible by the development of better ceramics and more efficient insulators for spacecraft.

The technology is simple. Off-peak electricity passes through elements to become heat energy. This raises the temperature of high-density ceramic bricks deep inside the radiator. Efficient insulation keeps the energy within these blocks overnight, after which it is released into the room through louvers at the top. Fans and dampers manage the output through preset timers and thermostats, which also control the heat input to the radiator. An attractive slim metal casing surrounds the whole, and they are available in a range of sizes. A prominent US manufacturer is Steffes, who can supply further up-to-date information (www.steffes.com).

Keeping Cool: Ice Storage

Heaven, for most Brits, would be to live in the southern states of the US where keeping cool is more of a problem than keeping warm. In these states the peak electrical demand comes from high air-conditioning loads. Ice storage is one potentially more economic and environmental alternative to this reliance on air-conditioning. The principle is to produce ice at off-peak rates and use it to release cold air later into a building. Grossmont Hospital in California has such a system installed; see www.calmac.com/downloads/Hospital.pdf.

For most householders, pre-cooling the house at cheap rates and then turning down the air-conditioning during peak periods is a crafty way of saving a dime or two. But don't overdo it, because you could end up with condensation and associated dampness problems.

THE CHOICE OF A HEATING OR COOLING SYSTEM is a complex matter because of the range of options available. It is difficult to cull general principles from the wealth of possibilities, but the following three are offered:

1. Fuel preference should be established first (see Chapter 4), since this determines running costs, environmental impact, and future availability.
2. A planned system is better than one that simply utilizes a number of individual appliances.
3. The decision should be based on informed choice.

I hope that this chapter has gone some way toward enabling you to make an informed choice. Tables 6.2 and 6.3 (pages 157–60) are offered as a summary of the various types of heating and cooling systems. Box 6.1 (page 161) gives tips to help you run your system efficiently.

Table 6.2 Heating Systems

System type	Fuels available	Brief description	Advantages	Disadvantages
Hydronic	Oil, gas, or solid. Electrical exists but is expensive to run.	Hot water is pumped through pipes to heat emitters such as baseboard convectors or radiators.	Suitable for any kind of property Pipes can be hidden. Integral hot water system Easy zone control	Eventual corrosion Possible leaks Emitters can be obtrusive. Slow response to changing temperatures Furniture positioning can be restricted.
Hydronic (ceiling or floor)	As above	Warm water is pumped through pipes embedded in the floor or ceiling.	Virtually invisible Easy zone control Integral hot water system	Not easily retrofitted (floor systems) Slow response time (floor systems) Control relies on predicting next day's temperatures (floor systems) Cold areas under tables, etc. (ceiling systems)

(continued)

Table 6.2 Heating Systems, *continued*

System type	Fuels available	Brief description	Advantages	Disadvantages
Forced air	Oil, gas, or electrical	Warm air is blown through ducts to controllable registers.	Heating and cooling in one system No corrosion or leakage Air can be filtered. Quick response to temperature change	Noise and odor transmission Convected heat only Separate domestic hot water system needed Zone control difficult New buildings only
Electrical (ETS radiators)	Electricity only	Heat energy is stored in radiators during off peak periods and released as required.	Easy installation Easily zoned Low initial cost	Convected heat only Separate hot water system needed Obtrusive radiators
Electrical (underfloor)	Electricity only	Heating wires are laid in concrete floor. Current taken at off-peak rates and heat is stored.	Invisible Easy zone control	Generally new buildings only, though element-embedded floor tiles are available Control relies on predicting next day's temperatures Separate hot water system needed

System	Energy source	Description	Advantages	Notes
Electrical ceiling heating	Electricity only	Heating wires are fitted inside the ceiling. Full-rate electricity is used.	Rapid warm up Invisible	Generally new buildings only, though element-embedded ceiling tiles are available Separate hot water system needed Possible cold areas under tables, etc.
District heating	Coal, oil, or natural gas	Hot water is supplied from a central boiler and distributed through pipes to individual buildings.	Lower environmental impacts Less "plant" in individual buildings Relatively cheap Otherwise, same as hydronic heating	Not generally available and requires high level of cooperation to establish a scheme Otherwise, same as hydronic
Combined heat and power, or CHP (large scale)	Electricity	Hot water cogenerated with electricity by local power station and distributed as in district heating	As in district heating	As in district heating
Combined heat and power, or CHP (micro)	Natural gas	Hot water and electricity provided by small household unit	Expected to be same as large-scale CHP	Cooperation between homeowners not required Still being test marketed

Table 6.3 Cooling Systems

System type	Fuels available	Brief description	Advantages	Disadvantages
Central air-conditioning	Electricity	Fan blows cool air through ducts into each room	Humidity control available Air can be filtered. Units can be reversible to produce warm air in winter.	As "forced air" in table 6.2
Room air-conditioning	Electricity	Fan blows cool air into room	As in central air-conditioning	As in central air-conditioning More than one unit needed for a house
District cooling	Electricity	Cold water extracted from a nearby lake	As in district heating	As in district heating Available in few areas

Box 6.1. Top Tips for Your Heating and/or Cooling System

1. Keep it clean! Clean or replace filters as needed—usually monthly. Clean registers, baseboard heaters, radiators, AC units, etc., and ensure that they are not blocked by furniture or drapes.

2. Replace old equipment. If your boiler or furnace is more than 15–20 years old, it is likely to be cost effective to replace it. Check the AFUE (annual fuel utilization efficiency) rating of any new equipment that you consider.

3. When replacing a boiler or furnace, consider either the more energy efficient condensing types or, better still, heat pumps.

4. Switch to district heating (or combined heat and power), if available. Watch for micro-CHP units becoming available.

5. Have your equipment professionally serviced annually. Often the amount it costs can be recovered in lower fuel bills.

6. Zone your system so that the bedrooms are separate from the living areas.

7. Balance the airflow. If some rooms are too hot and others too cold, adjustments are needed to rebalance the system.

8. In a hydronic system, bleed (or have bled) the radiators at least annually.

9. Fit radiator reflectors between radiators and walls.

10. Defrost your AC unit regularly or, if you have auto-defrost, periodically check that it is working.

11. Check that you are on the right rate plan. Use time-of-use (TOU) rates for electrical floor heating and thermal storage radiators.

CHAPTER 7

The Human Touch

Managing the System

Dad would have hated being called a good manager. He was a lifelong trade union activist, and "management" was the enemy. I didn't understand why until I got a vacation job working in the labs of his company. Then it was all too obvious. The company practiced "command management," and it was a classic case of management without leadership. Dad, like many of his generation, lost six years of his youth in World War II being pushed around and told what to do. After that he was determined that he wasn't going to put up with anything like it ever again. His union involvement was driven by a need to assert the dignity of working people.

Yet Dad was a good manager. He managed our finances well. We were never well off, but we always had clothes on our back, food in our belly, and a roof over our heads. What's more, we had a week's holiday by the sea every year, and Dad provided all I needed for my education. He was a good manager of our heating too, for without management, even the best system can leave you too hot or too cold and can be prohibitively expensive. When we eventually had central heating installed, Dad took to it with relish, altering this and that to see if he could get better performance. (Not always successfully, I might add!) In terms of functional management, albeit within a limited field, Dad was as good as they come. I often witnessed him deploying his resources in shrewd strategies in order to achieve his objectives—the basis of all good management.

For HVAC systems, the main resources are the various parts of the system (the *plant*) together with the fuel (the material) it uses. In a modern system there need be little labor involved, other than refueling, and ongoing finance is only required for upgrading, maintenance, and the purchase of fuel. The main objectives are to achieve and maintain comfortable temperatures and humidity in all occupied rooms whenever the building is in use, and to do so at maximum efficiency (i.e., at an economical cost with minimum manual intervention).

These objectives are met by using control equipment, which can be preset and left to operate our defined strategies quietly in the background. In other words, we use control equipment to make the system do as it's told.

Obedience Training

Brits are a nation of dog lovers, and I am no exception. When we first got our current dog Charlie (named after Charlie Chaplin because of how he walks), we soon discovered he was headstrong and difficult to control. Maybe it was because he was a mongrel, crossbred many times to give us a lovable, medium-sized pet and companion. He was strongly territorial and defensive of our household, his "pack." So we took Charlie to dog training classes. I was amazed to see the kinds of equipment available to help owners control their dogs. The equipment seemed to range from the hopelessly ineffectual to the downright cruel. So we didn't bother with any, but instead relied on the techniques the obedience trainers taught us. We soon learned that it is as much the owners that need training as the dogs.

When it comes to heating control equipment, the owner also often needs training before he can manage it effectively. For simple systems, the instruction leaflets supplied with the devices are usually sufficient, but as systems become more complex, a measure of "obedience training" may be needed.

But first let's start with a review of available control equipment.

Thermostats

Thermostats are the earliest and the best-known pieces of control equipment. This is not surprising, since the very first temperature control device was invented way back in the 1620s—by Cornelius Drebble in Holland as a new feature to his chicken incubator. Like virtually all thermostats, it depended on expansion, the increase in size when temperatures rise. Or as Mr. Hughes, my first physics teacher, put it, "When things get hotter, they get bigger."

Drebble's thermostat relied on the expansion of mercury in a tube. If the temperature became too high, its expansion pushed a metal bar to close a flap, which then restricted the airflow to the fire, causing it to die down a little. The process was reversed when the incubator cooled too much. The process is summarized in figure 7.1. Adjusting the temperature to different levels was possible by replacing the metal bar with a longer or shorter one. Ingenious!

Such was the genius of Drebble that there was little improvement in thermostats until a Brit, Andrew Ure, invented the *bimetal thermostat* some two hundred years later. These also rely on expansion, but of two different metals fixed together to form a strip. When the temperature rises, one metal expands more than the other, and the strip bends. This movement cuts off, or brings on, an energy supply.

Ure received a British patent in 1830, but his device wasn't used much until Professor Warren Johnson, a teacher at the Wisconsin State Normal School concerned about his students' comfort, used it as a classroom thermostat in 1883. He went on to found the still thriving Johnson Controls, Inc. The Honeywell Company can also trace its origins back to roughly the same time, when Al Butz combined Drebble's chicken incubator idea with Ure's bimetal strip to produce a "damper flapper" mechanism to control furnaces. A bimetal plate operated a motor that opened or closed a flap, thereby adjusting the air supply to the furnace. Smart thinking!

A "damper flapper" mechanism works reasonably well for solid fuel furnaces, but a thermostat directly operating a simple switch or valve can just as effectively control today's natural gas, oil, or electric furnaces in a home. This operates alongside a centrally located, bimetal wall thermostat, which operates the fan that blows warm air through the ducting. This gives a much more immediate response than controlling the furnace alone.

Wall thermostats respond to the temperature immediately surrounding them, which makes their positioning crucial. In fact, they are usually on an inside wall, so they are never in direct sunlight, which would make them "think" the air is hotter than it is and switch off the heating too soon. Fitted

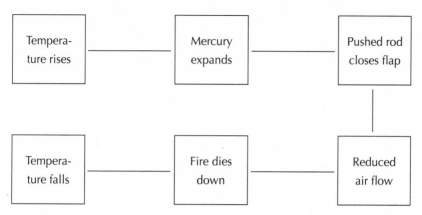

FIGURE 7.1.   Operation of Drebble's thermostat. The process is reversed on cooling. Modern thermostats work on the same principle.

like this they are also less affected by infiltrating cool air, which would have the opposite effect.

My friend Jim once asked me to see if I could fix his thermostat, for he had to set it at 80°F to keep his room at 70°F. I was glad to do so, for he had done me many favors over the years. But first I pointed out that although I knew the theory of heating well, I was no engineer. Luckily his problem was obvious. He had placed his desk under the thermostat and was running a computer and desk lamp under it. The rising warm air (convection again) was affecting the thermostat. He moved his desk a few feet along the wall and all was well.

Another problem is that if a wall thermostat is set to give a pleasant down-stairs temperature, upstairs can get pretty darn hot—convection yet again. Separate ducting is sometimes installed for different zones in the house, each individually controlled by its own thermostat.

So, Drebble's original damper flapper idea is still useful for controlling warm air systems. It wouldn't be much help with hydronic systems. In these, the wall thermostat operates the circulating pump, with a further thermostat to control the boiler temperature. As with warm air systems, the wall thermostat gives some control but is not ideal. What is needed is some form of automatic zone control, but before we tackle this, there is one further issue.

Strange though it sounds, the water in a boiler must not boil. If it does,

FIGURE 7.2.   Automatic zone valve

the water suddenly expands to steam of a much larger volume. This could blow up the boiler. The boiler thermostat is a vital piece of safety equipment, so an annual maintenance check by a qualified engineer is essential with a hydronic system.

Not all thermostats are bimetal strips. Wax pellets are also widely used. They control automobile cooling systems, as well as the automatic zone valves fitted to radiators in hydronic heating systems (see Chapter 6). A wax pellet expands as the temperature rises. This movement pushes a rod into a hole that cuts off the flow of hot water. It's not quite a damper flapper, but the idea of cutting off a flow is retained. This is the kind of engineering my dad would have loved: simple, cheap, and highly effective. These types of thermostats make automatic zone control possible. They are cheap enough for each radia-

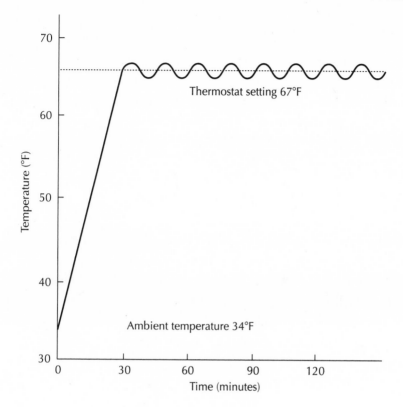

FIGURE 7.3.   Temperature response of a thermostat

tor to have its own, giving easy temperature control in every room of the house. Figure 7.2 shows one attached to my bedroom radiator.

One final point about thermostats. They are on-off devices, which continuously cycle on and off in response to temperature changes once the set temperature has been reached. Figure 7.3 shows the temperature change in a home in the morning as the building warms up. The human body cannot detect the small variations due to the cycling of the system, so they are not noticed and comfort is not compromised. Not so the warmup time—in this case thirty minutes—a cold bedroom would definitely be noticed first thing in the morning!

The warmup time varies tremendously according to the power output of the heaters, the level of insulation, the ambient temperature, and so on—but adjusting these factors alone would not give enough control to reliably and consistently achieve comfort. The best approach is to use timers in conjunction with thermostats to bring the heat on before the occupants get up in the morning. Correct adjustments of both timers and thermostats should eliminate the experience of a building being cold on account of its warmup time.

If keeping cool is more your problem, then exactly the same principles apply to the temperature responses of your cooling equipment, though in this case we talk about "cooling down" time.

Timers

I wonder what the pattern of heat use in your home is? Now that I am semiretired and at home most days, we run the heating all day in winter (and boy does it cost us!). Earlier in our lives, when Irene and I were both out working, it was only on for mornings and evenings—much more affordable! Automatic timers save the bother of continually having to switch the system on and off manually and allow for such different patterns of use. As you might expect, they operate the switches or valves controlling the fuel supply to the heating system. They are preset to override room thermostats so that the thermostat doesn't turn the system on when it should be off—overnight, for example.

Timers are easy to understand and operate. The basic timer is an electrical clock with a dial that goes around carrying little pegs. These click tiny switches on or off as they pass over them. The pegs are movable to different positions on the dial, so the functioning of the system can be matched to your pattern of use. They work well, but there are two problems. The first is that they don't function during a power outage, so they have to be reset. The second is that the pegs require a lot of fiddling to manipulate—not much use if you have arthritic fingers!

There's not much you can do about outages, but modern digital timers overcome the fiddliness. These have LCD screens (like a calculator) and various buttons to set on and off times throughout the day: much easier than manipulating small pegs. More sophisticated digital timers allow different settings for different days of the week, useful for those of us who go out on Saturday night and then sleep in on Sunday.

Timers can even be designed to allow different weekly patterns for different times of the year, so they only need to be set when installed. But just because it can be done doesn't necessarily mean it should be. We might wish for a stable climate pattern year after year, but it doesn't always happen, especially with climate change throwing in yet another uncertainty.

Programmable ("Setback") Thermostats

It would be silly to locate the thermostats and timers needed for optimum indoor comfort at different points in a building. Few people would enjoy running from room to room to reset everything. Modern technology allows everything to be set from one conveniently located digital display. This is made possible because *thermistors* have replaced bimetal thermostats. Thermistors are clever pieces of electronics whose electrical resistance, and therefore the current passing through them, varies with temperature, making it possible to display the temperatures on the LCD panel as well as controlling them remotely. Such a "box of tricks" can also include a sophisticated electronic timer, so "set and forget" really becomes possible.

These programmable thermostats, sometimes called *setbacks*, are becoming increasingly sophisticated. For example, sensors located outdoors can turn on the system before any temperature change is detected inside. Setbacks are also particularly useful if premises are left unoccupied. An elderly member of our church congregation "winters" in the Peak District. Her real home is on Skye, a rugged island off the coast of Scotland. The winter "up there" is too cold for her these days, so from about October to April she lives "down here." She has had controllers fitted to her Scottish home to prevent burst pipes and dampness due to condensation while she is away.

Humidity Control (Humidistats)

If you live by the sea (or a large lake) in a warm climate you will know how uncomfortable it can be. My early career visit to Dar es Salaam taught me that. The extra humidity keeps sweat from evaporating easily, making it difficult to get comfortably cool — not to mention the smell of stale perspiration making you unwelcome among your friends.

Indoor humidity, as briefly mentioned in Chapter 6, is controlled by humidistats. Like thermostats, these operate switches, but they do so in response to changes in humidity rather than temperature. Inside a humidistat there's a membrane (usually nylon) and a micro switch. The membrane shortens as it dries, and this operates the switch, turning on a humidifier. The membrane lengthens with increased humidity, causing a different switch to turn on a dehumidifier. Ingenius! Even more sensitive and reliable versions based on modern microchip technology are now becoming available. The chip continuously calculates the humidity and activates a switch when the humidity is outside the set range.

With humidistats, as with thermostats, there is a time delay in reaching comfort levels; removing excess humidity or introducing extra water vapor cannot be done instantaneously. Similarly, a timer should be used to keep the humidistat from operating when it is not needed.

Working Together

For relaxation I love to watch soccer on TV, especially if England is playing. The team is drawn from English players who play for clubs all over Europe. Getting such a bunch of proud individualistic players to work together must be very difficult. This is the role of the coach. Heating systems are similar, in that individual parts need to work well together to give optimum results. Getting this to happen is the task of the system's designer and requires every bit as much skill as building a soccer team. Incorrectly wired or connected components can cause the components either not to work at all or to interact in unexpected ways.

Even in an ordinary domestic system, it is surprising just how many controls there can be. Table 7.1 summarizes their role.

One final point about control. A few years ago, I was shown around a beautiful new set of apartments designed for the elderly. They were fully equipped with the latest electrical heating systems, air-conditioning, and neat wall-mounted controllers.

I spoke to the manager of the development a year or two later, and he had hit a major problem. Many of the tenants were simply too scared to touch the controls for fear of breaking them. The original settings had been left unchanged, with the heat on throughout the summer, the tenants controlling

Table 7.1   Control Equipment

Control equipment	Function
Boiler thermostat (hydronic systems, some hot water systems)	Controls the temperature of the water in the boiler
Room thermostat	Controls room temperature. Sometimes one is placed in a central area to give a measure of temperature control over the whole building
Automatic zone valves, or thermostatic radiator valves (TRVs)	Control the flow of hot water to a radiator in a hydronic system, thereby controlling the temperature of the room
Cylinder thermostat	Controls the temperature of the water stored inside the hot water cylinder
Frost thermostat	Overrides all other controls to turn on the system whenever there is danger of freezing
Outside temperature sensors	Turn on the system whenever the outside temperature falls to a predetermined level
Timer	Controls the times at which the heating and hot water systems operate
Programmable thermostat	Combines the function of a thermostat and timer in one instrument
Humidistat	Controls humidity levels
HVAC controllers	Like programmable thermostats but also control humidity

temperature by opening and closing windows. They simply had no idea how to use the system, and indeed were afraid of it. The manager was considering whether to send someone around twice a year to make the adjustments.

Any system is only as good as the person operating it!

Cunning Plans

Certain characters in the entertainment industry are held in high esteem on both sides of the Atlantic. The American Woody Allen springs to mind as a great talent held in high regard by both the public and the industry. On the British side, I suggest that Rowan Atkinson falls into the same category. His portrayal of Mr. Bean in *Bean: The Ultimate Disaster Movie* was hilarious to say the least. He also played the evil Emile Mondavarius in the 2002 *Scooby-Doo* movie with great professionalism, as well as the lead character in the spoof spy thriller *Johnny English*. In the UK he gained his fame on a TV series as Lord Edmund Blackadder. He and his sidekick—his dirty, smelly servant Baldrick (played by Tony Robinson)—were almost as well known and loved as Laurel and Hardy were in their generation. Baldrick had a "cunning plan" for every problem or difficulty. It was usually ridiculous but contained the germ of an idea that often helped get Blackadder out of trouble—often at Baldrick's expense.

Heating systems also need to be operated according to "cunning plans" if the overall aim of the greatest economy and minimum environmental impact is to be achieved. The optimum plan varies with lifestyle choices, but all such plans are based on the same underlying scientific principles. I have always found these plans easiest to understand by reference to motor vehicles.

Suppose you went on a journey in an automobile. If you drove hard across the US in an Aston Martin sports car (I wish, I wish!), it would cost you serious money. By comparison, a gentle run to a local picnic spot in your ordinary family car is a good cheap afternoon out for the whole family. The gas you use depends on how far you go, the model of the car, and how hard you drive it to reach your destination. In a similar way, the energy a building uses depends on how far you want the temperature to be above or below the outside temperature, the characteristics of the building, and how "hard" the heating system is used. This can be summed up in the following formula:

$$\text{Estimated annual energy use, } E = (T_1 - T_2)\, Bt \text{ BTU,}$$

where T_1 is the average temperature inside the building and T_2 is the average temperature outside the building, both temperatures being in °F, so $(T_1 - T_2)$ is how much the internal temperature is above (or below) the outside temperature. B is the building factor, which corresponds to the kind of vehicle (building) you have, and t is the length of the heating season in hours analogous to how hard the system is used.

Now, suppose that you move from Miami to a similar house in Seattle. You know it's going to cost you an arm in heating bills to live there, but will it cost you a leg as well?

Let's take a comfortable average internal temperature of 68°F as the temperature at which we set our thermostats. The mean outside winter (October to March) temperature for Miami is around 64°F and for Seattle it's around 45°F (data derived from http://gears.tucson.ars.ag.gov/beepop/cities.html). So the respective temperature differences are 4°F and 23°F, giving a temperature *difference* more than five times more for Seattle. Therefore you will use a little over five times as much energy in Seattle for heating and will emit correspondingly more carbon dioxide.

With this figure in front of you, you might wonder if you can afford the move—until you realize that you won't have any cooling costs in Seattle, which you had in Miami. The good news is that you (or your contractor) can use the same formula for estimating cooling costs. So:

Energy usage for Seattle = 5 times heating costs Miami − Cooling costs Miami

A further problem is that the new house will not be exactly equivalent to the old; the building factors will be different. The building factors depend on numerous things, in particular the level of insulation within the building, its volume, and its exposed external area. There is also a range of minor factors including orientation (does it face south?), exposure (is it on top of a windy hill?), and fenestration (how big are the windows?). However, provided you are not trading up to a much larger house of a vastly different construction and style, you can be fairly sure that the building factors are similar enough for an initial rough estimate.

But what about "cunning plans?" Again using the example of our automobile journey across the US, one "cunning plan" could be to drive in a less "hard" way, by getting the average speed down from the 65 mph typical on the Interstate system to a gentler 40 mph or so using side roads. The equivalent for heating a house is to develop plans that get the average internal temperature down. You could do this by simply lowering the thermostat setting, but that could also jeopardize your comfort. A better way is to ensure that your timer brings the heating system on only when it is required. The cunning plans below ensure that lifestyle patterns relate to system settings and show what savings can be made.

Granny's Plan

As my grandma got older, she felt cold more and more and kept her heat on day and night. In retrospect, we can see that she was wise, for hypothermia is a very real danger for older folk. Her plan was simply to maintain an average temperature of 68°F in her home by setting the controls to give this fixed temperature at all times. It is worth noting that other people, such as the very young, the very thin, and those with heart disease or circulatory problems may also be at risk of hypothermia and therefore may need to adopt this plan.

Figure 7.4 is a time-temperature graph for Granny's continuous plan over a twenty-four-hour period. The internal temperature is constant at 68°F. The external temperature is assumed steady at an average of 40°F, and the formula given earlier may be directly applied as follows:

$$\text{Estimated energy use, } E_1 = (T_1 - T_2)\, Bt = 28\, Bt \text{ BTU,}$$

where E_1 is the estimated energy use for case 1 (Granny's plan).

Students' Plan

Many students share a house with their friends. They share the rent, heating costs, and any other expenses involved, but it is not quite the same as a fraternity or sorority house, for they often move on as new friendships are formed and broken. For most students, a cunning plan would involve heating the

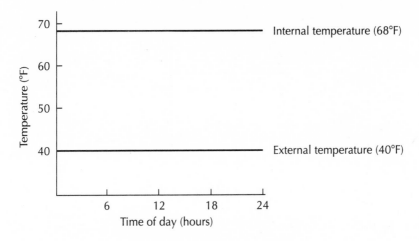

FIGURE 7.4.   Simplified time-temperature graph for continuous heating
(small temperature fluctuations ignored)

building for the daytime only. They pop in and out all day as they go to and
from lectures, but at night they would (or is it should?) be asleep in bed.

The students' graph is shown in figure 7.5. The heating system maintains a
mean internal temperature of 68°F, but only between 7 am and 11 pm. The
external temperature is unaltered at its mean of 40°F. The average internal
temperature over the whole day is calculated as follows:

8 hrs @ 40°F = 320
16 hrs @ 68°F = 1088
TOTAL = 1408
Thus, average internal temperature T_1= 1,408/24 = 58.7°F.

An alternative way of finding the mean temperature for the day is to use a graph
(figure 7.5). A line has been drawn on the graph parallel to the x-axis at the
temperature of 58.7°F. This divides the graph in a significant way. The area of
the part of the graph above the line is equal to the area of the part of the graph
below it, as shown by the shading. This gives a useful way of finding the mean
temperature when the graph is irregular. Draw a line that divides the graph into

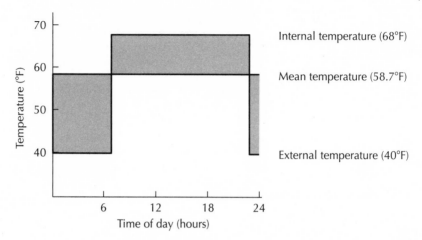

FIGURE 7.5.   Simplified time-temperature graph for daytime heating (small temperature fluctuations ignored and immediate warmup and cooldown assumed)

two equal sections in this way, and you have found the mean temperature.

Applying the formula now gives:

Estimated energy use, $E_2 = (T_1 - T_2)\ Bt = (58.7 - 40)\ Bt = 18.7\ Bt$ BTU,

where E_2 is the estimated energy used for case 2 (the students' plan).

The following can be used to compare the energy savings of the students' plan with that of Granny's plan *without needing to know the values of* B *or* t:

Percentage saving = $\dfrac{\text{original estimated energy use} - \text{new estimated energy use}}{\text{original estimated energy consumption}}$

$$\times\ 100 = \frac{E_1 - E_2}{E_1}\ \times\ 100$$

$$\frac{28Bt - 18.7Bt}{28Bt}\ \times\ 100 = \frac{9.3Bt}{28Bt}\ \times\ 100 = 33.2\%$$

Leave It on Overnight?

Many people wonder whether it's best to leave the heat on overnight. After all, if you turn the heat off, you save heat. Or do you? Even the most heavily insulated house will lose heat overnight, causing its temperature to drop and requiring reheating the next morning. So, what really matters is whether the heat saved by turning it off at night is more than that needed for reheating the next day.

The plans above give a clue to the dilemma. In Granny's plan, the heat is left on overnight. The student plan switches it off and reheats next morning. From our calculations, it looks like a savings of up to 33.2% can be achieved by switching the heat off overnight. But remember, these are estimates within which there are a number of simplifications. My guess is that you would be lucky to achieve a 25% savings in practice. Nonetheless, scientists who specialize in this kind of thing can show (using much more sophisticated equations) that there almost always will be a financial savings from turning the heat off overnight.

Working Families' Plan

When we first got married, both my wife and I went out to work. The house was unoccupied for much of the day, and the heating system was only needed for a short time in the morning and a longer time in the evening before being shut down for the night. Our heating plan is shown in figure 7.6.

The heating system maintained a mean internal temperature of 68°F between 7 am and 9 am and then again between 5 pm and 11 pm. The mean external temperature is, as before, 40°F. The mean internal temperature is calculated as before:

$$8 \text{ hrs @ } 68°F = 544$$
$$16 \text{ hrs @ } 40°F = 640$$
$$\text{TOTAL} = 1{,}184$$

Thus, mean internal temperature $T_1 = 1{,}184/24 = 49°F$, and the estimated energy use

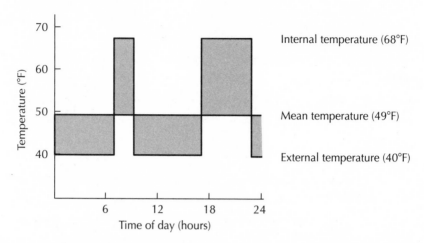

FIGURE 7.6.   Simplified time-temperature graph for morning and evening heating (small temperature fluctuations ignored and immediate warmup and cooldown assumed)

$$E_3 = (T_1 - T_2)Bt = (49\text{–}40)\, Bt = 9Bt$$

This gives percentage energy savings of 68% compared with continuous heating, and 52% compared with daytime heating only.

Warmup Times

The above three scenarios never actually occur in practice, although the calculations give useful rough and ready figures. One reason they are not realistic is that they assume instantaneous temperature rises as soon as the system comes on, which is impossible to achieve in practice. In other words, no account is taken of warmup times, which makes all the figures above underestimates. Figure 7.7 shows a more realistic situation for the student plan. The heating system comes on at 5 am in order to warm the building for use at 7 am, and it goes off as soon as the tasks for the day are finished at 11 pm.

In this situation it is difficult to find the mean internal temperature numerically, but a graphical method can be used. Look at the lines representing the average temperature in figures 7.5 and 7.6. The area of the shaded part of the graph above a line is equal to the area of the shaded part below a line. Drawing

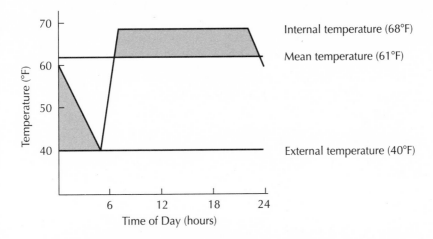

FIGURE 7.7.   Time-temperature graph for daytime heating, taking into account
warmup times (small temperature fluctuations ignored)

a similar line on figure 7.7 gives an average of 61°F, as opposed to the 58.7°F
of the less realistic figure 7.5. This gives a revised energy use of 21 *Bt* instead
of 18.7 *Bt*, meaning the original figure was around 11% too low.

Cunning Use of Warmup Times

As a young working family, my wife and I found it difficult to make ends meet,
especially when we were expecting our first child. She gave up work and we
readopted the "student" daytime heating plan. I realized that by delaying the
time at which the heating system came on, I could save money without sacri-
ficing comfort very much. I set the system to come on at 6 am and reach the
full comfort temperature of 68°F at about 8 am, but the internal temperature
was a tolerable 64°F at 7 am when we began our daily activities. The start of
the day is a time of physical movement around the house, so the slight decrease in temperature wasn't noticeable.

In exactly the same way, I realized that if I set the system to turn off an hour
earlier, at 10 pm, the house would cool down only a degree or so by the time
we went to bed at 11 pm. I reckon I saved around 6% throughout the heating
season, reducing both energy costs and environmental impact.

Other Plans

A further energy-saving modification, which I completely missed until I saw it in a book for heating engineers, is to increase the output power of the heating system by replacing the boiler with one of greater output. But how does this help? At first sight it looks as if much *more* energy is going to be used, since the output is so much higher. The secret is that it needs to be on for a much shorter time to bring the building up to its comfort level, because less energy is lost to the outside of the building. It's a bit like pouring water into a leaking bucket: the faster you pour when filling it up, the less water you lose. The shorter warmup time also means that the heating system can be turned on even later in the morning, another savings.

Other cunning plans include altering settings monthly to account for seasonal changes, fitting control systems that respond to outside temperatures, wearing extra clothes, and reducing internal thermostat settings. The effects of these plans can be quantified if the right monitoring equipment is available. One way of doing this is to carefully position *thermocouples*, a kind of electrical thermometer that automatically feeds the temperatures into a computer system at preset time intervals, within the building. The computer program then generates the graphs and calculates the average internal temperature. Building research institutes use this technique regularly (e.g., the Oak Ridge National Laboratory, at www.ornl.gov/sci/eere/buildings/index.htm, or for the UK, the Building Research Establishment, www.bre.co.uk/).

Research into these plans indicates that individually they tend to have a small effect (less than 5% reduction), but taken together they can make an important contribution to reduced costs and environmental impacts. These other plans are likely to become more widespread in the future, particularly as fossil fuels become scarcer and more expensive. We mustn't forget, however, the need to consider acceptable comfort and convenience alongside the need for economy.

The Bottom Line

It's always a good idea to bounce your ideas off a few friends before committing yourself to a course of action. When I showed a draft of this chapter to a few buddies, their response was, "Yes but what about the bottom line. How much should my bills be?" The obvious retort is, "Well look at your bills and apply the percentages." But in some circumstances this isn't enough. A neighbor of mine (a graphic designer) lives in a virtually identical house to ours. When we were talking one day, it came out that his electricity bill is about three times ours. So he went off and had a quarrel with his supply company. When they investigated, they eventually found that an electric fire had been left switched on in his attic for goodness knows how long. Oops! But unless he knew roughly what his bills should be, how would he know that anything was wrong?

Once you can estimate your fuel bill, not only will you avoid errors like this, but you can go a step further and repeat the calculation to see what effect changes are likely to have. This is a tremendous aid to decision making. Estimating a fuel bill requires the concept of design heat loss, and we turn our attention to that now. But a word of warning. The math of it is a bit tedious. So if math is a problem for you, you could easily leave this section out. There are a number of online tools or software packages that can be used instead. See, for example, http://hes.lbl.gov and yosemite.epa.gov/oar/globalwarming .nsf/content/ResourceCenterTools.html.

Design Heat Loss

Design heat loss (DHL) is simply the rate of heat loss from a building or a room within a building. It is measured in BTU/hr, the unit for the rate of flow of energy. It is made up of two parts, the fabric loss and the ventilation loss. If you think of a building as like a human body, the fabric loss is the heat energy lost though the skin, and the ventilation loss is that lost through the warm air you breathe out.

The size of the fabric loss depends on the surface areas of the building elements, their respective U-values, and the internal and external temperatures of the building. The concept of U-value has been discussed in Chapter 3, from which it follows that the rate of loss from an individual element (such as a wall or window) is:

$$\text{Rate of heat flow} = U \times A \, (T_1 - T_2) \text{ BTU/hr},$$

where U is the U-value in BTU/hr ft^2 °F; A is the exposed surface area in ft^2; T_1 is the higher temperature (usually that inside of the building) in °F; and T_2 is the lower temperature (usually that outside of the building) in °F.

Finding the total fabric loss for the building is therefore a matter of calculating the loss for each individual element and then adding them up. An alternative, which is mathematically equivalent, is to use the formula:

$$\text{Fabric loss} = (U_1A_1 + U_2A_2 + \ldots)(T_1 - T_2) \text{ BTU/hr,}$$

where U_1, U_2, etc. are the U-values of element 1, element 2, etc., and A_1, A_2 are their respective areas.

The *ventilation loss* is the rate of heat energy lost by escaping warm air. So the level of "draftiness" of the building is important. This is measured as the number of air changes per hour (ACH) experienced by the building. It generally varies from 0.25 to 1, but in very drafty buildings it can rise to as much as 2 or 3.

Ventilation loss can be calculated by using the specific heat of air. As air is a gas, this is defined most usefully for our purposes as the heat energy needed to change the temperature of unit volume (say 1 ft^3) by one degree Fahrenheit. Therefore:

Heat loss = volume of air × specific heat of air × its temperature change.

In symbols this becomes:

$$\text{Heat loss} = Vs\,(T_1 - T_2),$$

where V is the volume of air, here taken as the internal volume of the building (the volume of fixtures and fittings are generally ignored), s is the specific heat of air, T_1 is the higher temperature (i.e., that inside of the building), and T_2 is the lower temperature (i.e., that outside of the building).

If we multiply the heat loss by the number of air changes per hour, then the rate of loss per hour is obtained:

$$\text{Ventilation loss} = VsN\ (T_1 - T_2)\ \text{BTU/hr},$$

where N is the number of air changes per hour.

Since the specific heat of air (by volume at atmospheric pressure) is 0.018 BTU/ft³ °F, this simplifies to

$$\text{Ventilation loss} = 0.018\ VN\ (T_1 - T_2)\ \text{BTU/hr}.$$

And finally, adding fabric loss and ventilation loss gives the design heat loss:

$$
\begin{aligned}
\text{Design heat loss} &= \text{fabric loss} + \text{ventilation loss} \\
&= (U_1 A_1 + U_2 A_2 + \ldots)(T_1 - T_2) + 0.018\ VN\ (T_1 - T_2) \\
&= (T_1 - T_2)\{(U_1 A_1 + U_2 A_2 + \ldots) + 0.018\ VN\}
\end{aligned}
$$

since all the terms inside the curly brackets relate to a specific building, the formula can be further simplified to:

$$\text{Design heat loss} = (T_1 - T_2)\ B\ \text{BTU/hr}$$

where B is the building factor, discussed in the section on cunning plans earlier.

Estimated Annual Fuel Cost

Had I tried to take my graphic designer friend through the above math, I'm sure we would have given up at about the second stage and simply proceeded on the basis of the discrepancy in the bills. My cousin's daughter, Jane, however, recently moved into her first home and needed to know roughly what her heating bills might be. Guess who got the job of working it out for her!

Example 7.1 _____

The house has the following characteristics. What is its DHL and building factor?

1. The external walls have R-value of 15, giving a U-value of $1/15 = 0.067$ BTU/hr ft² °F and an area of 1,500 ft².

2. The attic has an R-value of 40, giving a U-value of 1/40 = 0.025 BTU/hr ft² °F and an area of 1,200 ft².

3. The basement and crawlspace both have an R-value of 19, giving a U-value of 0.053 BTU/hr ft² °F and also an area of 1,200 ft².

4. There are 0.5 air changes per hour.

5. The required average internal temperature is 70°F.

6. The average external temperature through the heating season is 40°F.

7. The internal volume of the house is 12,000 ft³.

Using the formula, the design heat loss

$$= (T_1 - T_2)\{(U_1 A_1 + U_2 A_2 + \ldots) + 0.018\ VN\}$$
$$= (70 - 40)\ \{(0.067 \times 1,500 + 0.025 \times 1,200 + 0.053 \times 1,200) + 0.018 \times 12,000 \times 0.5\}$$
$$= 30\{(100.5 + 30 + 63.6) + 108\}$$
$$= 30\ (302.1)$$
$$= 9,063\ BTU/hr$$

The building factor is 302.1, and the DHL is 9,063 BTU/hr.

The first stage was to find the design heat loss for the particular building. Example 7.1 shows how I did it, but in practice, energy managers, consumer advisers, and so on, set up a spreadsheet in a computer program. This has the added advantage that the effect of changing variables (e.g., adding cavity wall insulation) can be more easily explored. Table 7.2 shows what such a spreadsheet table might look like. I have based the calculations on a small single-story home, using recommended R-values for upper Midwestern states obtained from www.powerhousetv.com.

Finally, since the design heat loss is the rate of heat energy being lost from a building, it is also the rate at which heat energy must be supplied to maintain the internal temperature. Multiplying it by the number of hours in a heating season therefore gives the energy required per annum. This is converted to an estimated annual fuel cost by multiplying by the cost/useful BTU of the fuel being used to give:

Estimated annual fuel cost = DHL × hours in a heating season × cost/ useful BTU

Table 7.2   Spreadsheet for Determining Design Heat Loss

Surface	A	U	$T_1 - T_2$	Loss
Walls	1,500	0.067	30	3,015
Attic	1,200	0.025	30	900
Basement/crawlspace	1,200	0.053	30	1,908
Total fabric loss =				5,823
	vol	ach	$T_1 - T_2$	
Ventilation loss = 0.018	12,000	0.5	30	3,240
Design heat loss				9,063
Building factor =				302.1

Example 7.2 _____

What is the estimated annual heating cost for the house in example 7.1? The heating season is 2,500 hrs, and oil is used at a cost/useful kWh of 6.3 cents. What are the savings if a natural gas system is installed at a cost/useful kWh of 3.6 cents? (1 kWh = 3,412 BTU.)

For the oil, the cost/useful BTU is 6.3/3,412 cents = 0.00185 cents.

Estimated annual fuel cost   = DHL × hours in a heating season × cost/useful BTU

= 9,063 × 2,500 × 0.00185 cents

= 41,916 cents

= $419.16

For the natural gas, the cost/useful BTU is 3.6/3,412 = 0.00106:

Estimated annual fuel cost   = 9,063 × 2,500 × 0.00106 cents

= 24,017 cents

= $240.17

Thus, the estimated annual savings is $419.16 – $240.17 = $178.99

The average external temperature during the heating season, used in calculating the DHL, can be found from past meteorological data. The number of hours in a heating season depends upon the patterns of use of the building. The cost / useful BTU can easily be derived from the costs / useful kWh in table 4.4. Example 7.2 applies all this information to the same house as in example 7.1.

There are, of course, various uncertainties in the data given for the house and heating season. For this reason, I would quote an annual cost in the range of $400–$450 and savings in the range of $150–$200 per year.

How Big Is Beautiful?

A few years ago, I was asked to attend a meeting of the property committee of the Methodist church I attend. They were considering a new scheme that involved building an addition to house a new small meeting room and bathrooms with handicapped-accessible stalls. They wanted to know whether to fit a new boiler for the existing hydronic system, or whether its output could cope with the extra demand. I took the line that big enough is beautiful, too big is expensive, and too small leads to people feeling cold and complaining.

So I set about my calculations to find the output capacity needed, and as I was in the UK I worked in kW. I calculated the design heat loss as 30,000 W, or 30 kW, so it looked as if a boiler of that size should be enough. There was a problem though. The design heat loss is based upon an average external temperature for the heating season. Yet, the system must be capable of keeping the building at the internal design temperature even on days on which the external temperature is less than the average. In practice, the external temperature is taken (in the UK) as $-1°C$ (the industry standard), and the design heat loss must be recalculated with T_2 set to $-1°C$. Recalculating in this way gave me a figure of 32 kW. Since available equipment comes in standard outputs, the next largest available boiler — 35 kW — was chosen and has proved satisfactory.

When applying these calculations to your home system, remember that both output and input capacities are usually quoted for heaters. Be careful not to confuse them. Also, if domestic hot water is to be supplied from the same unit, extra capacity must be added.

You could also work on a room-by-room basis to see if the output of each room's heaters is large enough. And if you really get into it, you can do before and after calculations to estimate savings for the various measures you have in mind!

Home Energy Audits

I have been unable to trace the origin of the well-known saying, "The whole is greater than the sum of its parts." It used to puzzle my mathematical brain, until I realized that it was not about math at all. It came to me in a flash as I was cooking a delicious mixed vegetable soup using the produce of my own labors from our garden. Eating the soup was much better than eating the ingredients separately. Then I began to see other applications of the principle. Your car will not stop properly if you only service the brake system and ignore the tires. The point is that all the bits interact, and changing one can affect the others. Sometimes this can happen in unpredictable ways, but more often the combined effect is far better than what might be expected from carrying out each improvement on its own.

A holistic approach is necessary, one that treats the building as a whole. A well-conducted home energy audit does exactly that. At its simplest, such an audit consists of applying the tips at the ends of the chapters of this book to your own home in a systematic way. If you wish to do this, first customize these tips to your property by constructing a checklist for your own system. Then go around on a tour of inspection, noting what things need to be adjusted and what is due for upgrading. You can then either upgrade on a do-it-yourself basis or call in the appropriate professionals.

However, it is well worth having your home professionally audited from time to time. The auditor will go into much greater detail, doing a room-by-room examination of your home and checking your utility bills against what a reasonable expectation of your energy consumption might be. To do this, the auditor will need to know such things as your average thermostat settings and the patterns of use of the building. So have this kind of information readily available for their visit to your home.

A professional audit should also include a *blower door test* and a *thermographic scan*. The blower door test establishes the building's airtightness to establish if existing air sealing is adequate. This requires specialist equipment and takes about an hour. A thermographic inspection uses infrared cameras to produce visible images of the temperature of the building's walls, windows, and other surfaces. From this, the pathways by which heat escapes from the building can be seen, and the temperatures can be compared with what would normally be expected. Thermographic scans are particularly useful in determining the effectiveness of the building's insulation.

More detailed advice on energy audits can be obtained from the US government website www.eere.energy.gov/consumer/your_home/energy_audits/index.cfm. It is certainly worth considering.

A HEATING SYSTEM CAN BE MANAGED manually switching it on and off at appropriate times, but this is inconvenient. Control devices are more efficient. There are three main control devices: thermostats, humidistats, and timers. Thermostats respond to temperature changes, and timers bring the system on or switch it off, according to their settings. A system may need more than one thermostat to control the temperatures of its various parts. Humidistats control humidity in much the same way that thermostats control temperature. Control systems are becoming so sophisticated that central programmable thermostats can be fitted. The human touch remains essential, however, as programmable thermostats still have to be appropriately set and adjusted.

But what is appropriate? This depends very much on the lifestyle of the occupants of the household and can vary from continuous heating to using the heat for just a few hours in the morning and a few at night. "Cunning plans" can be devised to minimize energy use while maintaining acceptable levels of comfort.

It is also possible to calculate the bottom line for any situation using the concept of design heat loss. The effect of changes to the system can then be explored using before and after calculations. Sizing of equipment also matters—outputs must not be too low or you will end up cold. On the other hand, if outputs are too high, the equipment will be unnecessarily expensive.

Finally, as any engineer will tell you, a system is only as good as the person operating it!

Box 7.1. Top Tips for Managing Your System

1. Find your own comfort levels:
   - Adjust your heating thermostats to the lowest comfortable temperature in winter. Heating costs are reduced by about 2% for every 1°F reduction in the settings.
   - Adjust your cooling thermostats to the highest comfortable temperature in summer.
2. Install a programmable thermostat and use it to adjust your system to function according to your lifestyle.
3. Adjust your thermostats and timers to take account of warmup and cool-down times.
4. In hydronic systems, fit automatic zone valves to all your radiators.
5. Carry out an energy audit based on the tips at the end of the chapters of this book. Consider having a professional audit done.

CHAPTER 8

The Crystal Ball

A Look at the Future

The arrival of a traveling fair to a neighborhood promises thrills and excitement. When I was a boy, a traveling fair visited our town twice a year. We went together as a family, and I have fond memories of riding the bumper cars with Dad and of Uncle Derek winning prizes at the shooting gallery. Aunt Dora's specialty was to take me to see Gypsy Rose Lee to have my fortune told—until it eventually dawned on us that her predictions were nothing more than broad generalities of little practical use.

In this chapter, I too indulge in a little futurology. My musings are not based on images in a crystal ball like Rose's but on what is already happening today, for I believe, with the French philosopher Simone Weil, that "the future is made of the same stuff as the present." Nonetheless, I am well aware that predictions can turn out hopelessly wrong. I first became environmentally concerned back in the late seventies and still have an authoritative book on my bookshelf that predicted that the world would run out of oil by the year 2000, or 2010 at the outside. Well it hasn't happened, and a lot of the Greens from that time have been left with egg on their face.

So I am cautious, but I still believe that the thinking at that time was correct, though the time scale was wrong. The earth really *does* have finite resources of fossil fuels, and they *will* eventually run out. What no one realized back then was that global warming was an even greater threat to our world. Current thinking is that global warming is real, man-made, and likely to render the earth uninhabitable long before the fuels run out, on account of the carbon dioxide they release when they burn. Not that it is going to happen the way it does in Roland Emmerich's film *The Day after Tomorrow*. The film suggests that global warming could trigger an abrupt shift in our planet's climate. The story focuses on the problems of a young man, Sam Hall, trapped in New York City as he grapples with plunging temperatures and severe flooding amid the mass evacuation of people to the South. However, the consensus among scientists is that dramatic shifts like this just won't happen. Nonetheless, the effects of global warming pose some very real threats.

So I conclude that a major driver of change in the way we use energy, in addition to normal economic forces, will be the threat of global warming. Previous chapters of this book contain useful information about what we can do individually to help counter the threat, without making ourselves uncomfortable and miserable. But, these tips are based on today's technology and today's availability. In this final chapter, I hazard a guess at future innovations.

Current equipment will continue to be improved and become more widespread. This is particularly true of conservation measures such as insulation (Chapters 2 and 3) and control systems (Chapter 7). Greater use of heat pumps and combined heat and power (Chapter 6) will also occur, as will the further adoption of renewable resources for power generation. There will also be better burner designs and more efficient furnaces and boilers, though dramatic improvements seem unlikely, for we have gone as far as currently conceivable with these "old" technologies.

Most people will, however, continue to purchase the best available conventional equipment they can afford. This alone will improve the situation, as today's "kit" will be cleaner and more efficient than yesterday's. Some will vote with their energy dollars for cleaner/greener electrical supplies, as their awareness of environmental issues rises and as more suppliers offer customer packages that include electricity from renewable resources. Pace University, of White Plains, New York, has already developed a website (www.powerscorecard.org/) that enables consumers to rate specific electricity suppliers according to their environmental impact as well as by price. This kind of initiative will continue to grow.

Small Is Beautiful

According to the *Times Literary Supplement* of October 6, 1995, the book *Small Is Beautiful: Economics as if People Mattered*, by E. F. Schumacher, ranks among the hundred most influential books published since World War II. Though born German, Schumacher made his home in Britain and became very influential in the US—so much so that his library is housed in South Egremont, Massachusetts. In her essay "E. F. Schumacher, an Appreciation" (Hannum 1997), Nancy J. Todd notes that as early as 1995, Schumacher said, "A civilization built on renewable resources, such as the products of forestry and agriculture, is by this fact alone superior to one built on non-renewable resources, such as oil, coal, metal, etc. This is because the former can last, while the latter cannot last." Is this merely prophetic idealism? Perhaps, but this position has a much stronger foundation than anything Gypsy Rose might adhere to and is gaining more credence with every passing year.

Some people accept Schumacher's position and, seeing that steady improvement of existing technology will not be enough, are entering the field of *micro–power generation:* their homes are being redesigned and equipped to generate at least some of their power needs from renewable resources. This is not so new an idea as it first appears. When Thomas Edison built the first power plant in Pearl Street, Manhattan, in 1882, he had a vision of delivering power close to where it was used through a system of local "central" stations that served relatively small districts. At the same time, others were installing generators in their factories and homes. Indeed, in 1907, 59% of America's electricity came from small-scale generation of this sort.

Today's microgeneration equipment has been gradually developed over the last thirty years or so. Yet problems inhibit its widespread use. The first is economic. "Normal" fossil fuel equipment is, at present, cheaper to buy and to run, though the alternatives are rapidly catching up. The second problem is that equipment based on renewable energy sources needs appropriate sites—there is little point in installing a wind generator in the middle of the Nevada desert or solar panels in Alaska. The third is the intermittent nature of the renewables; the sun doesn't shine at night and the wind doesn't blow consistently and predictably.

Nonetheless, microgeneration is slowly gaining ground, and there are groups of enthusiasts dedicated to it. Their magazine *Home Power* contains some excellent well-informed articles covering the whole field, and their website has some first-rate free introductory downloads (www.homepower.com/index.cfm). If you think, however, that microgeneration and alternative fuels are the province solely of the enthusiast, then think again. Major oil companies are investing in these technologies (see www.shell.com/renewables), and BP has renamed itself *Beyond* Petroleum, instead of its old title British Petroleum. These actions surely speak for themselves.

I am therefore confident that these technologies could become mainstream in the not-too-distant future. Experience is already being gained, and the market is steadily growing. My look at the future in this chapter focuses mainly on microgeneration utilizing renewable fuels.

Wind

Wind turbines harness the energy of a moving mass of air to rotate a shaft, which then turns an electrical generator. The electricity produced is generally DC (direct current, i.e., battery electricity). This is not too much of a problem for general lighting using ordinary light bulbs, but it doesn't work with fluorescent-tube lighting (or low-energy lamps, which are based on tube technology). For most household equipment an *inverter*—a small electronic box—is needed to convert the DC to AC (alternating current, or mains electricity) before it can be used extensively in the home. Fortunately, these devices are reliable and not too expensive.

The smallest available turbines are a few hundred watts and are widely used by the boating community to charge their batteries while at sea. Larger 1.5 to 6 kW versions can produce a more useful quantity of electricity for households. These are placed either on the roof or on a freestanding mast in the garden, since wind speed (which determines output) increases with height. The visual impact (and sometimes noise) can lead to problems, and the installation may contravene local state or city codes, so laws and codes should be checked beforehand. To make turbines worthwhile, a locality must have sufficient wind, though it doesn't need to blow either steadily or continuously. Meteorological data may be available to assess this; otherwise long-term measurements must be made. One company marketing wind-power equipment is www.windsave.com.

The unpredictability of the wind is summed up by a verse in the Bible: "The wind blows where it chooses, and you hear the sound of it, but you do not know where it comes from or where it goes" (John 3:8). Fortunately, the changing direction of the wind is easily dealt with by incorporating a "weather cock" mechanism to always turn a turbine's blades to face the wind. A bigger issue is the intensity of the wind, which can change from minute to minute. One solution is to have the generator charge up a bank of batteries and draw a current from them via an inverter when needed. A backup diesel generator can be incorporated for use when wind speeds are low and the batteries flat. Equally, there has to be a way of "shedding" excess electricity when the batteries are already fully charged. Diverting it to space heating or water heating can usefully do this.

There is, however, a better way, but it requires the agreement of the local electricity supply company. This is simply to connect to the electricity grid and sell excess electricity to the company when you have an oversupply and buy from it when you have a shortage. The necessary equipment for this is available today; some of the latest simply use a dedicated plug-and-socket system to connect directly to the grid.

When I was a young boy in the early 1950s, the must-have consumer appliance was naturally the TV. It was easy to tell who had one because of the large antenna fixed to the highest point on their house. It became a powerful status symbol for a short while, telling us something about the wealth and priorities of our neighbors. Could small wind generators eventually have the same sort of impact, I wonder? Imagine how much fossil fuel would be saved if every home were a generator.

There is, however, an upper limit on what is achievable this way. Varying amounts of energy added to the local grid at unpredictable times puts a strain on control systems. If more than 20% or so of electricity were supplied to the grid system like this, it could become unstable, leading to outages. Nonetheless, we are a long way from the 20% figure, and currently there are no techni-cal problems involved.

Enthused? There is authoritative and independent advice on the US De-partment of Energy's website at www.eere.energy.gov/windandhydro/wind_consumer_faqs.html and www.eere.energy.gov/consumer/your_home/electricity/index.cfm/mytopic=10880. The American Wind Energy Association (www.awea.org/) offers additional detailed information.

Solar

Passive Solar Heating

Just before I retired, I treated myself to a large conservatory on the south-facing wall of our bungalow. It is great for sitting in during spring or fall, for we can enjoy the warmth of the sun even when the outside air temperature is rela-tively low. I also keep a few tomato plants in it during summer, so we can eat them fresh with our evening meal. It acts as a huge greenhouse, and if I keep the door connecting it to the house open, warm air infiltrates the rest of the house. I therefore need to burn less fuel to keep warm. Since the building does nothing but receive this warmth, it is known technically as *passive solar heat.*

Architects in recent years have given much thought to passive solar design. It is no accident, for example, that atria form an important design feature of many new homes. Houses are also consciously being orientated to face the south (in the northern hemisphere), to maximize the solar energy gain. An-other less obvious "trick" is to have large windows on the south-facing side of the house and smaller ones on the north side; have a look at any newly built houses in your neighborhood to see if this has been done.

Many of these design features are relatively cheap to incorporate in a *new* house, but less so in one already built. Altering window sizes is disruptive and expensive and may not be allowed in your neighborhood. It probably wouldn't be permissible if you lived in a historic colonial home, for example.

Most buildings in my village were built two hundred or more years ago out of the local gritstone, and the quarries are still visible (and a hazard to the unwary) in the hillsides surrounding our valley. The dense walls of these older homes warm up during the day, and in the evening slowly release their energy into the building, thereby saving fuel. They are effectively a short-term energy store for solar heat, and such a feature can be designed into modern buildings.

In the late 1950s, a French engineer named Felix Trombe developed this simple technology in his wall designs. A typical Trombe wall consists of eight- to sixteen-inch-thick masonry, which has been coated with a dark material to increase the absorption of solar radiation. Glass is placed three-quarters to two inches in front of the wall to trap the incoming heat energy—the green-house principle once again. More sophisticated designs have a fan that passes air slowly over the wall to speed up the heat extraction in the evening. The wall must, of course, face south, and they are only effective and economic in climates with a fair number of hot days followed by cool nights. The National Renewable Energy Laboratory (part of the US Department of Energy) has more details on its website www.nrel.gov/.

Passive Cooling

Passive cooling aims to protect the building from the heat of the sun and sur-rounding air by using nonmechanical methods, saving energy and cooling costs. When I returned from my time teaching in Uganda in 1970, I soon became (temporarily) hooked on watching TV in the evenings. This was something I had missed while away from home. By the late seventies, I had become more discerning in my taste, and one of the programs I looked out for was *Dallas*. Week after week, like many others, I followed the ups and downs of the Ewing family. Of particular interest to me was the construction of their home—the South Fork ranch—which had many features that encour-aged passive cooling. I notice now that the house is a major tourist attraction, though I sadly doubt that I will ever get there. I must content myself with the pictures on the Internet.

Passive cooling techniques are used in climates where the main issue is day-time cooling. Because many of the techniques are the exact opposite of those needed for passive heating, building designs cannot be optimized for both pas-sive heating and passive cooling. For example, passive heating emphasizes the use of dark colors on the outside of the building (to absorb radiant heat), whereas passive cooling emphasizes light colors (to reflect it). South Fork is a good exam-ple of this. Similarly, in passive cooling main living areas are positioned within the building to protect them from direct sunshine in the summer, whereas the exact opposite layout can be advantageous in colder climates.

Maximum use is also made of shading. Planting vegetation such as trees or large shrubs is a way of doing this, as is the placing of architectural features such as overhangs and awnings between the sun and the house. A shaded veranda around the building is a useful architectural device for providing shade: verandas pretty much surround South Fork, both upstairs and downstairs. Shutters or roller blinds can also be used to some advantage, but, of course, opening and closing them is an extra job for the home's occupants.

Ventilation strategy also varies with climate. Far from attempting to reduce them, the aim is to make full use of available breezes. An open construction allows air both to enter a building and move easily from room to room. However, insulation, as described in chapters 3 and 4, is just as good at keeping heat out as it is at keeping it in, so don't go down the route of doing the exact opposite in everything.

Much, of course, depends on the local climate and the particular characteristics of the site. Nonetheless, I would expect passive heating and cooling techniques to be more extensively used in the future, and we should all know a little about them. Inspired? More details about both passive solar heating and passive cooling are available at www.eere.energy.gov/RE/solar_passive.html.

Solar Water Heating

In passive solar heating, the Trombe wall acts like a greenhouse: the glass allows in short-wave energy from the sun, and the energy is then absorbed by the wall, warming it up. Glass won't let out the longer-wave energy re-emitted by the wall, so energy is trapped within the structure, slowly warming it up. Solar water-heating panels work in the same way. At their simplest, a sheet of glass covers a thin box through which water trickles, emerging warmer than the temperature at which it entered the panel. A further refinement involves painting the inside of the box black to help absorb the heat of the sun. Apart from the panels, a hot-water storage cylinder and some simple pipework are needed.

My wife Irene and I recently had a short vacation in Cyprus. To our surprise, virtually every house had its own solar water-heating panel (figure 8.1).

FIGURE 8.1.   Solar water-heating panels on a house in Cyprus

Both the climate and the already high fuel prices on the island make solar panels competitive with all other alternatives. Alas, solar panels are of little use to me. The British climate is variable to say the least, and the figures just don't stack up. The problem isn't so much the capital outlay (a basic system can be installed for around five thousand dollars), but the long payback times of ten years or more. Mind you, as fossil fuel prices start to rise, this could drop dramatically, and it's worth reviewing the situation from time to time. Details are available about the use of solar panels in the US from the Department of Energy: www.eere.energy.gov/RE/solar.html and www.eere.energy.gov/consumer/your_home/space_heating_cooling/index.cfm/mytopic=12490.

Solar Photovoltaics

My son despairs of me. My cell phone is five years old, and I've had my calculator for at least fifteen years. According to him, I am the most uncool father in town. I am particularly fond of my calculator, bought for me by my mom just before she died. The interesting thing about it is that, in spite of regular use, it has only once had to have new batteries.

The reason is that it is solar powered, or to be more accurate, solar assisted. There are four small solar cells known as *photovoltaics (PVs)* on the front. These generate electricity whenever *any* light—not just direct sunlight—falls on them, so the internal chemical batteries are only needed on the rare occasions when I use it in very poor light conditions. Consequently, they last for ages.

PVs also power satellites and spacecraft by converting solar energy into electricity just as ordinary batteries convert chemicals into electricity. They generate no climate changing greenhouse gases and are very long lasting, with low maintenance requirements. *Building-integrated photovoltaic materials (BIPVs)* serve the dual purpose of generating electricity and acting as a construction material. They can be fabricated into many different forms. Probably the most common form are panels that can be integrated into roofs, but they can also be used as part of a window or as external cladding.

With no moving parts, BIPVs are reliable devices with expected lifespans greater than thirty years, and with twenty-year guarantees now available. They are modular in nature, meaning that they can be added to or removed and used elsewhere. The modules are not expected to create any out-of-the-ordinary disposal problems at the end of their lives, for they are largely made of nontoxic silicon crystals, chemically equivalent to beach sand.

As with wind turbines, local permissions may be necessary before installing BIPVs, and they too produce DC electricity intermittently—even in the Sunshine State, Florida, there is no sunlight at night! This means BIPVs need either storage batteries or an import-export arrangement with the local grid. The set-up costs are currently high—around eight thousand dollars per peak kW—and a kW isn't that much power for many homes. Economic payback times are therefore long.

Prospects, however, look good. The current overall energy efficiency of the best cells is around 20% and rising, whereas not so long ago it was a mere 5%. As technology continues to improve, payback times will drop dramatically, and we may see some fossil fuel power stations closing as a result. Definitely a technology to watch closely. See www.eere.energy.gov/RE/solar_pho

tovoltaics.html for more technical details, and if you are really serious about installing your own system, see www.eere.energy.gov/consumer/your_home/ electricity/index.cfm/mytopic=10710.

Small-Scale (Micro) Hydropower

Hydropower has been around for a long, long time. The Greeks are known to have used waterwheels for grinding wheat into flour more than two thousand years ago, and here in the UK, waterwheels are part of our industrial heritage. They abound as fascinating tourist attractions. Using falling water to generate electricity didn't really take off until the 1880s. In 1882, the world's first hydro-electric power plant started generating from the Fox River in Appleton, Wisconsin, and by 1889 there were two hundred such plants in the US alone.

It is easy to see why hydropower took off so rapidly. The technology is relatively simple—falling water simply rotates a turbine (a more efficient waterwheel) connected to a generator. The energy efficiency of large schemes can be in excess of 90%, though for small home systems, 50% is more realistic. The rate of water flow and the height from which the water falls are critical in determining the power output and cost per unit of electricity generated.

It is surprising how small a stream can produce a useable amount of electricity—thirteen inches of water is enough for a small submersible turbine. If you have a big enough site, it is possible to filter and divert some of the river water through a channel to a turbine house for a much larger output. The reliability of the water flow may be an issue however—streams or rivers can dry up—as is output variability. Export to or import from a grid may also be necessary, as for wind and solar systems, to iron out fluctuations in output. Small-scale hydropower schemes therefore are highly site-specific in their feasibility and economics. Further details are available from www.eere.energy .gov/windandhydro/hydro_technologies.html and www.eere.energy.gov/con sumer/your_home/electricity/index.cfm/mytopic=11060.

Solid Smoke and the Zero-Energy Home

The dream of many environmentalists is to achieve a zero-energy home. In many ways, designing a zero-energy home is an extension of the whole-house approach; it certainly needs all relevant professionals to be intimately involved and working as a team. Such a home would optimize many of the design features described in this book such as passive solar heating and cooling, climate-specific design, and state-of-the-art insulation, to reduce demand. The challenge would then be to use the technologies described in this chapter to supply all the occupants' power needs, resulting in a net zero energy consumption from the utility provider.

This ideal would not have to lead to an uncomfortable caveman-style home. In fact, in many respects comfort and convenience would be improved. For example, a highly insulated building would mean fewer temperature fluctuations. Furthermore, a home that produces its own electricity is pretty much immune from outages in the grid system. It would also contribute to environmental sustainability by saving energy and reducing pollution.

Can it be done? Well, the technologies described in this chapter certainly work, and there is much promising research underway. As I write, NASA's Stardust probe has just returned safely to Earth, and a material called *aerogel* has made the headlines as the medium for "catching" stardust. Invented by Steven S. Kistler of the College of the Pacific in Stockton, California, way back in the thirties, aerogel is made from the same material as glass, namely silicon dioxide, which is extremely abundant (it's sand!) on the earth's surface. It is manufactured into a porous, sponge-like material one thousand times less dense than glass and many times better as a thermal insulator than the best fiberglass.

Imagine what this "frozen smoke" or "puffed-up sand" could do as an insulating material. Not only could it make zero-energy homes a real possibility for many people using existing technology, but it could also bring really low energy bills within the reach of everyone in the US. Alas, there are still problems with aerogel, mainly to do with manufacture. It is extremely fragile and needs some kind of framework to hold it together. Moreover, processing times are long, and it is difficult to achieve consistency in the quality of the end product, both of which make it costly to manufacture at present. Nonetheless, research is ongoing, and its future looks promising. At present the market leader seems to be Aspen Aerogels, Inc. of Northborough, Massachusetts (www.aerogel.com/).

Finally, the US Department of Energy has a research initiative underway with partner building professionals and organizations to build zero-energy homes. Although only a few prototypes are up and running at the moment, first indications look good. The future is not all doom and gloom by any means, provided we are sensible in our decision making. If you wish to follow up on the zero-energy idea, these websites will help:

> www.eere.energy.gov/consumer/your_home/designing_remodeling/index
> .cfm/mytopic=10360
> http://usgovinfo.about.com/cs/consumer/a/zeroenergy.htm
> www.davisenergy.com/zeh_page.html.

SO THERE IS MUCH TO GIVE US HOPE for the future. The history of technology shows that what is cutting edge today will be commonplace tomorrow. So, if you see yourself as an early adopter of new technology and have the funds to indulge your fancy, I hope this chapter has given you an overview of where things are now and what your future might hold.

As for my wife and me, we love the idea of living in a low-energy home but don't have the money to purchase a newly built house. So we will continue improving what we now have in as environmentally friendly a way as possible. But if innovation excites you, then I encourage you to go for it, for the risk is low compared with the potential rewards.

From Whence We Came

Although I enjoy indulging in a bit of nostalgia from time to time, I've never been a great one for looking back. We cannot do anything about the past. It's interesting, though, to speculate about what would happen if we could change it, and there are plenty of books and TV programs that have indulged in this kind of thinking. *Star Trek* episodes spring to mind, as does H. G. Wells's novel, later made into a film, *The Time Machine*. That, however, is the world of science fiction, and the real world we inhabit is rather different.

Historians tell us that we can learn from what has gone before and use it to plan a better future. I hope that this is where reading this book has left you, eager to plan a better future in the little bit of the world that is your home. Or if you are a student, I hope it has formed a stimulating overview of the field. Yet it will not be enough for more detailed planning or study; for that you will have to find a more comprehensive text or consult people more expert and experienced than I. Nevertheless, I hope that this book succeeds as a helpful and readable introduction, for that was its original purpose. It should be enough, at least, to get you started.

A key theme recurring through the book is that of environmental responsibility. This relates mainly to global warming, which is now pretty much accepted by the world's scientists as both real and anthropogenic in nature. Climatologists' computer models predict not just that everything will become a degree or so warmer, but also instabilities in the atmosphere (more and stronger hurricanes for example), thermal expansion of the oceans (leading to coastal flooding), and a myriad of other scary scenarios. The main culprit is thought to be the increased concentration of greenhouse gases, such as carbon dioxide and methane, in the atmosphere. This book emphasizes the considerable part we can all play together in heading off these effects.

Yet global warming is not the only concern. Waste disposal is also a worry, and we have touched on this too. Are the materials we use recyclable at the end of their life? If not, can we recover the energy within them by burning them, rather than dumping them in landfill sites? The source of the fossil fuels on which we remain dangerously dependent is also a problem. They *will* run out one day, and the Western world still relies heavily on imports from countries that may not always be friendly to us. Renewables have a part to play in any national strategy—and, as Chapter 8 shows, microgeneration systems based on solar, wind, and hydropower are available now.

But what about lifestyles? Does environmental responsibility mean that we have to revert to a caveman lifestyle? I hope that I have convinced you that this is not so. An intelligent application of current technology can give us *both* comfort and economy. These concepts are not incompatible, and indeed, many of the measures described here pay for themselves from the money saved by using less fuel and go on to show a "profit" after that.

In this final short chapter, I present you with a brief review of the journey this book has undertaken. The logical starting point, described in Chapter 1, is to think about what kind of environment we need to create within our homes to make our bodies feel comfortable. The right temperature is clearly one consideration. Luckily, we find that our bodies can happily adapt to a range of temperatures. Temperatures between 60°F and 75°F are usually fine, but it does depend a bit on what we are doing and wearing. We don't need as high a temperature if we throw a party involving lots of energetic dancing, for example. Humidity is another major issue, but the comfort range is broad, with 40%–70% relative humidity feeling pretty much acceptable to all. However, if too much of this humidity arises from the perspiration of our partygoers, our room could well finish up with odor problems. Apart from these factors, there are other less obvious issues, which we briefly covered, including temperature gradient and air movement (drafts). The purpose of doing all of this was to set out the conditions needed to achieve our aims.

Encouragingly, Chapters 2 and 3 described win-win situations. Insulation saves money, leads to less carbon dioxide emissions, *and* can increase comfort. There are, however, a bewildering array of insulating materials and techniques. On the other hand, there is no shortage of advice, and almost no one should have difficulty finding something suitable for their property once they know what they are looking for. R-values give an easy way to compare *materials*. Comparing *methods* is a bit more complex, because they vary according to the particular building element. You cannot cavity-fill your windows or double-glaze your walls. I did hear, however, of someone who brought in contractors to cavity-fill a solid stone masonry wall! Luckily, they didn't go ahead and only time was wasted.

Most books consider weather-stripping part of insulation. The other insulation techniques work equally well for keeping heat out and keeping it in, but in hot southern states you may wish to encourage air movement to allow cooling breezes to enter the building. Much depends on the particular climate, building, and depth of the owner's pockets.

We hear almost as much about problems of fuel supply these days as we do about the dangers of global warming. Hurricane Katrina and the Iraq wars have kept this issue prominent in our minds. Luckily, in most areas a choice of fuel for heating is available, so the impact of reduced supplies need not be a major crisis, provided we don't mind replacing or adapting the heating system we have. Yet there is more to fuel choice than availability. For most of us, cost is the crucial issue, and this was given serious consideration in Chapter 4. Working out the true cost of fuel is not as straightforward as it first appears, for we must take into account the different units in which the fuels are sold, as well as the different efficiencies of the appliances in which they are used. I hope the explanations I have given, and the examples worked through, straighten out this important matter. Further concerns of versatility, safety, convenience, and environmental impact were mentioned, too, to give a more complete picture of the situation.

Prior to researching this book, I never knew how much there was to the science of fire and fireplaces. Sure I knew the principles, but the detail turned out to be extensive and fascinating, particularly as it took me into so much American and British history. I hope this fascination came through in the text of Chapter 5. For wood and the fossil fuels, the technical issues are ease of ignition and safe and controllable burning. This requires an adequate air supply to the flames and a safe escape pathway (the flue) from it. None of these problems occur with electric fires and fireplaces, but to my mind, the electric versions are not as attractive. Aesthetics is important for devices that are on constant view. Once again there is a bewildering number of options. I hope this chapter has helped you find your way through this particular maze.

Fires and fireplaces, at best, heat only one room. For whole-house heating you could have, as I was brought up with, a fire in every room. A much more energy-efficient, convenient, and safe solution, however, is central heating (Chapter 6), which has been around since Roman times. Development was slow at first, and it wasn't until the twentieth century that central heating really took off. Blowing hot air from a furnace around ducts and into rooms is by far the most common system in the US. Hydronic systems, with water pumped from a boiler through radiators, are more common on the British side of the Atlantic and are used in the US too. Interesting energy-efficient variants of these two systems include the use of heat pumps to supply warm air, and of district heating or combined heat and power to supply hot water. Electric heaters are another viable option, and the amazing variety of different forms and models available reflects the flexibility of electricity as a fuel in the home. But, of course, for many people the problem is keeping reasonably cool rather than warm. So Chapter 6 also compared different types of cooling systems.

Whatever system you have, it must be controlled in some way. Imagine a system that pumped heat energy into a well-insulated house long after a comfortable temperature had been reached. Wow, would it get hot! So Chapter 7 covered control. Thermostats have been around since the 1600s and still remain an essential component of any system. These have now been supplemented with humidistats, timers, and programmable devices of one kind or another. The technology in general is very reliable, but as in so many fields, the same cannot always be said of its operators. In this chapter, the advantages and disadvantages of a few basic patterns of use were explored. Unlike other chapters, this one attempted to make quantitative estimates of cost through the concept of design heat loss. The calculations tend to be long and tedious rather than complex. Don't worry overmuch if you haven't followed them completely—there are plenty of professionals and websites that will do the job for you. Personally, I always find it satisfying to do a few examples of the "sums" from first principles, as here, before resorting to a computer program. You may be different, and not a lot is lost if you skip this bit.

Finally, I indulged in Chapter 8 in a little futurology. The emphasis was on microsystems, many of which are available now and, I (tentatively) predict, will become commonplace in the future. Appropriate use of small-scale solar, wind, and hydropower systems will both substantially reduce our dependence on fossil fuels and significantly lessen our carbon dioxide emissions. The ultimate goal is, however, a zero-energy house, which uses a combination of heavy insulation, renewable fuels, and sensible practices to achieve indoor comfort throughout all seasons. A bit far-fetched? Maybe. But remember prototype buildings are being built *now* to test this concept.

Most of us, however, have less ambitious aims than to turn our homes into zero-energy havens. We want to be warm and comfortable at the lowest possible cost in the home we already have, though we rightly worry about environmental issues. Whatever your aspirations relating to comfort, I hope that this book has gone some way toward helping you to meet them, and that it has been an interesting and informative read.

Appendix

Conversion Factors

Length

1 mile (mi)	= 1.609 km	1 kilometer (km)	= 0.621 mile
1 yard (yd)	= 0.914m	1 meter (m)	= 1.094 yd
1 foot (ft)	= 304.8 mm	1 millimeter (mm)	= 0.0394 in
1 inch (in)	= 25.4 mm		

Weight

1 ton (tn)	= 1016 kg	1 tonne (t)	= 1,000 kg
			= 0.984 tn
1 pound (lb)	= 0.454 kg	1 kilogram (kg)	= 2.205 lb
1 ounce (oz)	= 28.35 g	1 gram (g)	= 0.0353 oz

Area

1 square foot (ft^2)	= 0.093 m^2	1 square meter (m^2)	= 10.76 ft^2

Volume

1 gallon (US)	= 3.785 liters	1 liter	= 0.001 m^3
1 gallon (UK)	= 4.546 liters		= 61.03 in^2
			= 0.264 gal (US)
		1 cubic meter (m^3)	= 1.308 cu yd

(continued)

Temperature

Fahrenheit

temperature (°F) $= 32 + 1.8C$

Celsius

temperature (°C) $= 5/9(F - 32)$

Energy

1 British thermal unit (BTU)	$= 1,055$ (J)	1 Joule (J)	$= 0.2389$ cal
	$= 252$ cal		$= 9.5 \times 10^{-4}$ BTU
	$= 0.293 \times 10^{-3}$ kWh		
1 Therm (th)	$= 100,000$ BTU	1 kilowatt hour (kWh)	$= 3.6 \times 10^{6}$ J
	$= 105.5 \times 10^{6}$ J		$= 3,412$ BTU
1 calorie (cal)	$= 4.184$ J		
1 food calorie (Cal)	$= 1,000$ cal		
	$= 4,184$ J		

Power

1 horsepower (hp)	$= 745.7$ W	1 watt (W)	$= 1$ Js^{-1}
1 British thermal unit per hour (BTU/hr)	$= 3.412$ W	1 kilowatt (kW)	$= 1,000$ W

Thermal transmittance (U-value)

1 BTU per hour per square foot per degree Fahrenheit (BTU/hr ft^2 °F)	$= 5.675$ Wm^{-2}C^{-1}	1 watt per square meter per °C (Wm^{-2}C^{-1})	$= 0.176$ (BTU/hr ft^2 °F)

Glossary

ACH. An abbreviation for number of air changes per hour, used as a measure of the "draftiness" of a room or building.

AFUE rating. The annual fuel utilization efficiency of a piece of equipment, such as a furnace or boiler. The higher the better.

Albedo. A measure of the reflectivity of a surface. Often expressed as a percentage, it is the fraction of radiation striking a surface that is reflected by that surface.

Air barrier or air retarder. *See* breather membrane.

Air-conditioning. Equipment used to regulate the temperature and humidity in a building or room. Sophisticated systems also control the circulation and quality of the air.

Ambient. An adjective meaning "of the surrounding area or environment," as in "ambient temperature."

Boiler. A device that produces hot water or steam.

Breather membrane (or air barrier, air retarder, or housewrap). A membrane fitted around the outside of a house that allows water vapor to diffuse outward but prevents liquid water from getting in.

British thermal unit (BTU). The imperial unit of energy. One BTU is the amount of energy it takes to raise the temperature of one pound of water by one degree Fahrenheit. This is roughly the amount of heat energy given out by burning a kitchen match.

Calorie. A unit of energy most often used for heat energy. Originally the amount of energy it takes to raise the temperature of one gram of water from 15 to 16 degrees Celsius. NB: One food calorie is equal to one thousand calories (1 food calorie = 1 kilocalorie). *See also* Joule, kilowatt hour, and British thermal unit.

Calorific (heat) value. The energy available in one unit of fuel, used to compare fuels for the amount of heat energy they contain.

Carbon dioxide. A gas formed when fossil fuels (or wood) burn completely. It has been implicated in climate change.

Carbon monoxide. A gas formed when fossil fuels (or wood) do not burn completely. Highly poisonous.

Central heating. A heating system in which the heat energy is distributed from a "central" source to other parts of a building.

Chimney. A hollow vertical structure for carrying away waste gases and smoke. *See also* flue.

Chimney cap. A structure at the top of the chimney to prevent rain, downdrafts, birds, etc. from entering the flue. May need careful design in some localities. *See also* flue terminal.

Coefficient of performance (COP). The ratio of heat energy output to electrical energy input of a heat pump. Sometimes loosely referred to as a heat pump's efficiency.

Cogeneration. *See* combined heat and power.

Combined heat and power (CHP). Equipment designed to produce both heat energy and electricity from a single energy source. Originally called "cogeneration."

Condensate. Water originating from steam or water vapor that has condensed.

Condensation. The appearance of moisture (water vapor) on a surface, caused by warm moist air coming into contact with a colder surface.

Conduction (thermal or heat). The passage of heat through a body without large-scale movements of matter.

Conductivity (thermal or heat). A quantity that measures how well heat energy flows through a material by conduction.

Conservation of energy. A fundamental law of science that states, "Energy can neither be created not destroyed, only changed in form."

Contraction. The decrease in size (volume, length, or area) of an object due to a lowered temperature.

Convection (forced). The transfer of heat energy from one place to another by the mechanical movement of a gas or liquid. Fans or pumps push the warm fluid from place to place.

Convection (natural). A heat transfer process involving currents within a (fluid) liquid or gas, caused by differences of density within the fluid and the action of gravity. The differences occur because of expansion in the hotter parts of the fluid, which then rise.

Cyclical systems. A system in which a fluid is returned to its starting point before being sent off again in a cycle. Examples are warm air, steam, and hydronic heating systems. Refrigerators and heat pumps also rely upon a fluid passing around in a cycle.

Dehumidifier. A device for artificially decreasing the humidity in a room or building.

Density. A measure of how heavy a material is. The mass per unit volume of the material.

Design heat loss (DHL). The *rate* of heat energy loss from a building (or room within a building).

Direct vent. A furnace or boiler design in which all the air for combustion is taken from the outdoors and all exhaust products are released to the outdoors. Also known as sealed combustion.

District heating (DH). Heating systems that distribute steam or hot water to a number of buildings across a district.

Double glazing. Double-pane window with air space between the panes to provide additional insulation.

Duct. A pipe, tube, or channel for conveying gases, liquids, or solids from one point to another.

Efficiency. The useful output compared with the total input—usually expressed as a percentage.

Electrical resistance heater. A heater (e.g., electric baseboard) in which the heat energy is produced by an electric current flowing in resistance wires.

Embodied energy. The energy used in processing a product and transporting it to the place where it is used.

Emissivity. The ability of a surface to emit radiant energy compared to that of a black body at the same temperature and with the same area. Emissivity ranges from 0 to 1.

Energy. Ability (or capacity) to perform work. Scientifically, work is done only when a force moves an object. Measured in joules, calories, kilowatt hours, and British thermal units.

Evaporation. The change from a liquid to a gas below its boiling point, due to the escape of fast-moving molecules from its surface.

Expansion (thermal). The increase in size (volume, length, or area) of an object due to increased temperature.

Expansivity. A quantity that measures how much a material expands when its temperature is increased.

Fabric loss. The heat energy loss through the fabric (walls, windows, roof, etc.) of a building.

Flue. The passageway through which waste gases are removed from a room or building.

Flue terminal. A structure at the end of the flue passageway to prevent rain, downdrafts, etc., from entering the flue. May need careful design in some localities. *See also* chimney cap.

Framed construction. A type of construction in which a frame (usually wood) is erected prior to panels, windows, walls, etc., being fitted to the frame.

Fuel. Material containing one type of energy that can be transformed into another more useful form of energy. Can be thought of as an energy store.

Fuels (fossil). Fuels derived from fossils, including coal, oil, and natural gas. Can only be replaced on a geological time scale.

Fuels (renewable or alternative). Fuels that are capable of being replaced naturally within a short time relative to the earth's natural geological cycles. Include solar, wind, and tidal energy.

Furnace. A device used to heat air.

Gallon. A unit of volume applied to liquids. The US gallon is different than the UK gallon, which is 1.2 times larger. The UK gallon is 4.546 liters, and the US gallon is 3.785 liters.

Gravity system. A way of circulating air or water that relies on convection only.

Heat. Generally used in this book to mean a form of energy—thermal energy. A stricter, more accurate definition would be that heat is a measure of the amount of energy transferred from one body to another because of the temperature difference between them.

Heat exchanger. A device for allowing heat energy to move from one place to another efficiently.

Heat pump. A device that uses compression and decompression of gas to heat and/or cool a house. Can have very high efficiencies.

Heat transfer coefficient. See thermal transmittance.

Housewrap. See breather membrane.

Humidifier. A device for artificially increasing the humidity in a room or building.

Humidistat. A control device used to maintain humidity close to a preset point.

Humidity. The dampness of the air.

Humidity (absolute). A way of measuring the dampness of the air. The mass of water vapor per unit of volume of space. Measured in grams per cubic meter or grains per cubic foot. A grain is a unit of weight; 1 pound = 7,000 grains.

Humidity (relative). The amount of water in the air compared with how much the air can hold at the current temperature, expressed as a percentage.

HVAC. An acronym for heating, ventilation, and air-conditioning.

Hydronic system. A central heating system that uses hot water pumped through radiators.

Hygrometer. A device for measuring relative humidity.

Incidental gain. The heat energy within a building from a casual unplanned source such as body heat or heat from electrical equipment.

Infiltration. The uncontrolled inward leakage of air through cracks and gaps around a building, especially around windows and doors (the technical term for a draft).

Infrared radiation. See radiation.

Insulation (thermal). A material that slows down the flow of heat passing through it.

Isolation valves. Valves that isolate a piece of equipment from the supply (of oil, gas, or water) to it. If a radiator is leaking, closing the valve stops further water from entering it.

Joist. A beam used to support floor or ceiling loads. (Roof joists are sometimes called rafters.)

Joule. The fundamental metric or SI unit of energy. It is about a quarter of a calorie.

Kill switch. A security mechanism used to shut off a device in an emergency situation. *See also* thermal cut out.

Kilowatt hour (kWh). The common metric unit of energy, a kilowatt hour is the amount of energy transformed by a device of one kilowatt in one hour. (A domestic iron is approximately 2 kW.)

Latent heat. *See* specific latent heat.

Masonry construction. A type of construction in which stone, brick, concrete, or similar materials are bonded together with mortar to form the walls.

Metabolic rate. The rate at which the body uses energy.

Payback time. The time it takes to recover the initial outlay of capital.

Photovoltaic. A device that converts light energy from the sun into electricity.

Pound. (1) The imperial unit of weight equivalent to 0.454 kilogram. (2) The UK unit of currency.

Power. The rate at which energy is changed from one form to another.

Radiation. Energy transport through electromagnetic waves. For heat energy, the waves are in the infrared part of the electromagnetic spectrum.

Radiator. A device designed to emit infrared radiation.

Register. A grill that covers the opening of a duct in a heating or cooling system. Can usually be opened or closed to regulate the flow of air.

R-value. *See* thermal resistance.

Sealed combustion. *See* direct vent.

Setback thermostat. A thermostat with a timer that can be programmed to various temperatures at different times of the day/week. *See also* thermostat.

Shutoff valve. A valve designed to prevent flow within a piping system after it has been closed.

Solar gain. The contribution to the heating of a building made by the sun's energy entering it, for example, through a window.

Specific heat capacity. A measurement of how much energy is needed to heat up a material or of how much energy a material can store. In imperial units, it is the heat energy required to raise the temperature of one pound one degree Fahrenheit.

Specific latent heat. The heat energy that is either released or absorbed by a substance when it undergoes a change of state, such as during evaporation, boiling, or melting. In imperial units, it is the heat energy released or absorbed by one pound of the material when it changes state.

Temperature. The degree of hotness or coldness. Usually measured in Celsius (°C) or Fahrenheit (°F).

Therm. A unit of heat energy most commonly used in the natural gas industry. 1 therm = 100,000 BTU.

Thermal-cut out. A switch that shuts off an electrical device in danger of overheating. Unlike a thermostat, it stays off until reset.

Thermal resistance (R-value). A measure of the ability of materials to transfer heat energy. A low value means that a material is a good conductor, and vice-versa.

Thermal transmittance (U-value). A measure of how fast heat energy goes through a structure under fixed conditions. It is the rate of heat flow through unit surface area of a structure when opposite sides are in contact with the air at temperatures that differ by one degree. Also known as the heat transfer coefficient.

Thermometer. An instrument for measuring temperature.

Thermostat. A switching device for maintaining a predetermined temperature by switching the energy source on when the temperature is lower than the desired value and off when it is above the set value.

Ton. An imperial unit of weight equal to 2,240 pounds.

U-value. See thermal transmittance.

Vapor check (also called a vapor retarder). A material that restricts the rate of flow of water vapor through a building's outer envelope.

Vent. An opening that allows the flow of air through it.

Ventilation. The process of supplying air to or removing air from a space.

Ventilation loss. The heat energy loss from a building or room as a result of ventilation or infiltration.

Ventless. A gas fireplace design in which all the air for combustion is taken from indoors around the equipment, and all exhaust products are released indoors. Can be hazardous and must comply with current legislation and building codes.

Warm (forced) air system. A heating system in which warm air is blown through ducts into rooms.

Watt. A unit of power. A rate of energy conversion of one joule per second.

Work. The transfer of energy from one object or system to another by applying a force over a distance.

Further Reading

Bynum, R. T., and Rubino, D. (2000) *Insulation Handbook*. Hightstown, NJ: McGraw-Hill Education.

Cooper, Gail. (1998) *Air-Conditioning America: Engineers and the Controlled Environment, 1900–1960*. Baltimore: Johns Hopkins University Press.

Curtis, Peter S., and Breth, Newton. (2002) *HVAC Instant Answers*. Columbus, OH: McGraw-Hill Professional.

Gordon, J. E. (1968) *The New Science of Strong Materials*. New York: Walker and Co.

Hannum, Hildegard, ed. (1997) People, Land, and Community: Collected E. P. Schumacher Society Lectures. New Haven, CT: Yale University Press.

Hebra, Alex. (2003) *Measure for Measure: The Story of Imperial, Metric, and Other Units*. Baltimore: Johns Hopkins University Press.

Kingsley, Charles. (1863) *The Water Babies: A Fairy Tale for a Hand-baby*. London: Puffin.

May, Jeffrey C. (2001) *My House Is Killing Me! The Home Guide for Families with Allergies and Asthma*. Baltimore: Johns Hopkins University Press.

National Research Council. (1985) *District Heating and Cooling in the United States*. Washington, DC: National Academies Press.

Norgate, T. E., and Rankin, W. J. (May 2002) "The role of metals in sustainable development," Proceedings, Green Processing 2002, International Conference on the Sustainable Processing of Minerals: 177–184.

Parkinson, C. Northcote. (1957) *Parkinson's Law*. New York: Ballantine Books.

Reid, E. (1988) *Understanding Buildings: A Multidisciplinary Approach*. Cambridge, MA: MIT Press.

Schumacher, E. F. (1973) *Small Is Beautiful: Economics as if People Mattered*. London: Frederick Muller.

Silverman, Steve. (2001) *Einstein's Refrigerator and Other Stories from the Flip Side of History.* Kansas City: Andrews McMeel Publishing.

Simms, Andrew, et al. (2005) *Mirage and Oasis: Energy Choices in an Age of Global Warming.* London: New Economics Foundation.

Simon, Julian L. (1981) *The Ultimate Resource.* Princeton, NJ: Princeton University Press.

Stranks, Jeremy W. (2005) *The Handbook of Health and Safety Practice,* 7th ed. Harlow: Prentice Hall.

Thoreau, Henry D. (1854/1995) *Walden: Or, Life in the Woods.* Mineola, NY: Dover Thrift Edition.

Wilder, Laura Ingalls. (1935) *The Little House on the Prairie.* London: Egmont Books.

Wilder, Laura Ingalls. (1940) *The Long Winter.* New York: HarperCollins.

Wilson, Alex. (1998) *Consumer Guide to Home Energy Saving,* 6th ed. Washington, DC: Island Press.

USEFUL WEBSITES

Please remember that the contents of websites are constantly changing as they are updated. If the following addresses do not take you to the site of interest, you can usually find a good alternative by using a search engine.

www.ashrae.org/. The website of the American Society of Heating, Refrigerating, and Air-Conditioning Engineers. Has a useful consumer section.

www.awea.org/. The American Wind Energy Association.

www.baywinds.com/new/FossilFuel.html. An essay comparing the cost of small-scale wind power with fossil fuels. Some useful links.

www.bchydro.com/powersmart/. The Powersmart program of the British Columbia Hydro Company. Useful and relevant downloadable information.

www.calmac.com/downloads/Hospital.PDF. Based in New Jersey, CALMAC designs and manufactures off-peak cooling systems utilizing thermal energy storage, an innovative way to air-condition buildings.

www.cat.org.uk. The Centre for Alternative Technology (UK).

www.cpsc.gov/cpscpub/pubs/463.html. The US Consumer Product Safety Commission has factsheets on the safety of heating and air-conditioning equipment. This factsheet refers to supplemental room heaters.

www.customfireplaceandmore.com/faqs.htm. A supplier of fireplaces, etc., based in Cookeville, Tennessee. See also the New Jersey–based site www.mendhamfireplace.com/index.htm.

www.davisenergy.com/zeh_page.html. The Davis Energy Group is one of four national teams working to develop zero-energy homes.

www.diydata.com/. Lots of useful information for DIY householders. UK-based.

www.doe.gov. The home page of the US Department of Energy.

www.dow.com/. The website of Dow Chemicals, manufacturers of Styrofoam insulation.

http://www.eere.energy.gov/. The US Department of Energy's, energy efficiency and renewable energy site. Extensive and covers both technical and practical issues as well as the latest programs. Lots of downloadable consumer information.

www.eia.doe.gov/. The website of the US Energy Information Administration, the official source of energy statistics for the US.

www.energy.ca.gov/distgen/equipment/stirling_engines/vendors.html. The site of the California Energy Commission. This particular page relates to Stirling engines, but there are many more pages of interest.

www.energyshop.com/. A site for homeowners and businesses that compares electricity and natural gas prices.

http://energystar.gov/. Energy Star is a US government–backed program helping to protect the environment through energy efficiency. Lots of information and details about how to access help from the program.

www.est.org.uk/myhome/generating/types/wind/. The UK Energy Saving Trust has useful information on home generating and other energy-saving measures.

www.eweb.org/home/energy/index.htm. Oregon's largest public utility, the Eugene Water and Electricity Board has useful, intelligible, and relevant factsheets for the consumer.

http://gears.tucson.ars.ag.gov/beepop/cities.html. Part of the website of the Carl Hayden Bee Research Center, this page lists mean monthly temperatures for US cities.

www.hearth.com/what/specific.html. Part of a site specializing in stoves and fires.

www.heatinghelp.com/. An informational site with a name that speaks for itself.

www.homebuildingmanual.com/Glossary.htm. A glossary of construction terms.

www.homeenergy.org. The website of the magazine *Home Energy*, containing a lot of useful information for the homeowner. Past articles are archived and freely accessible. Well worth subscribing to the magazine.

www.homepower.com/index.cfm. The website of a magazine dedicated to home-scale renewable energy and sustainable living, particularly microgeneration.

www.idwr.state.id.us/energy/renewables/default.htm. The Idaho Department of Water Resources has good educational pages on its website.

www.naima.org. The website of the North American Insulation Manufacturers Association.

www.ncat.org/. The National Center for Appropriate Technology (US).

www.nrel.gov/. The US Department of Energy's national renewable energy laboratory. Lots of information for homeowners, teachers, students, and others.

www.omni-test.com/Publications.htm. The Omni Test Laboratories, Inc. website has many downloadable pages mainly concerned with wood fires, stoves, and air quality.

www.ornl.gov/sci/. The website of the Oak Ridge National Laboratory, the US government's largest science and energy laboratory, which has a strong energy efficiency and renewable energy program.

www.powerscorecard.org/. A website that allows electricity suppliers to be compared according to the environmental impact of their supplies.

www.shell.com/renewables. A site dedicated to what Shell Oil Company is doing regarding renewable energy sources.

www.steamlocomotive.com/. The steam rail enthusiasts' site for the US. (See also the UK equivalent, http://ukhrail.uel.ac.uk/facts.html.)

www.steffes.com. A US manufacturer of electric thermal storage equipment.

www.vermiculite.net/. A basic introductory page about vermiculite insulation.

www.windsave.com. A UK company that develops and sells wind turbine microgenerators.

http://yosemite.epa.gov/oar/globalwarming.nsf/. The US Environmental Protection Agency's global warming site.

Index